Intelligent Systems Reference Library

Volume 239

The aim of this series is to publish a Reference Library, including novel advances and developments in all aspects of Intelligent Systems in an easily accessible and well structured form. The series includes reference works, handbooks, compendia, textbooks, well-structured monographs, dictionaries, and encyclopedias. It contains well integrated knowledge and current information in the field of Intelligent Systems. The series covers the theory, applications, and design methods of Intelligent Systems. Virtually all disciplines such as engineering, computer science, avionics, business, e-commerce, environment, healthcare, physics and life science are included. The list of topics spans all the areas of modern intelligent systems such as: Ambient intelligence, Computational intelligence, Social intelligence, Computational neuroscience, Artificial life, Virtual society, Cognitive systems, DNA and immunity-based systems, e-Learning and teaching, Human-centred computing and Machine ethics, Intelligent control, Intelligent data analysis, Knowledge-based paradigms, Knowledge management, Intelligent agents, Intelligent decision making, Intelligent network security, Interactive entertainment, Learning paradigms, Recommender systems, Robotics and Mechatronics including human-machine teaming, Self-organizing and adaptive systems, Soft computing including Neural systems, Fuzzy systems, Evolutionary computing and the Fusion of these paradigms, Perception and Vision, Web intelligence and Multimedia.

Indexed by SCOPUS, DBLP, zbMATH, SCImago.

All books published in the series are submitted for consideration in Web of Science.

Seiki Akama · Yotaro Nakayama · Tetsuya Murai

Epistemic Situation Calculus Based on Granular Computing

A New Approach to Common-Sense Reasoning

Springer

Seiki Akama
Kawasaki, Kanagawa, Japan

Yotaro Nakayama
Biprogy
Kōtō, Tokyo, Japan

Tetsuya Murai
Chitose Institute of Science and Technology
Chitose, Hokkaido, Japan

ISSN 1868-4394 ISSN 1868-4408 (electronic)
Intelligent Systems Reference Library
ISBN 978-3-031-28553-0 ISBN 978-3-031-28551-6 (eBook)
https://doi.org/10.1007/978-3-031-28551-6

This Springer imprint is published by the registered company Springer Nature Switzerland AG
The registered company address is: Gewerbestrasse 11, 6330 Cham, Switzerland

Foreword

Artificial Intelligence studies nowadays mostly focus on machine learning techniques including deep learning and related research, while the fundamental aspect of the field remains within the framework of classical as well as non-standard logics. The latter logical framework moreover includes non-standard set-theoretical systems like fuzzy sets and rough sets. Thus, a discussion of the latter theories is needed to proceed with further studies of artificial intelligence in its fundamentals.

To cover all these broad topics is not easy for researchers, as they are mostly working in a narrow area, as they are pursuing deep fundamental theoretical properties of a logical system.

Fortunately, however, the three authors of this monograph succeeded in carrying out this difficult task: this book encompasses various logical systems, rough sets, and related subjects, while fundamental issues such as the results concerning soundness and completeness are rigorously described.

The main topic in this monograph is the granular reasoning of the epistemic situation calculus. Besides the logical framework of the situation calculus, the granular computing based on rough set theory enables the appropriate setting herein. The predicate of possibility is introduced and used throughout. Moreover, logical deduction systems using Gentzen type calculi and semantic tableaux calculi are introduced. These considerations finally arrive at discussing the Frame Problem, the central issue in artificial intelligence. Readers will find the granular computing treatment of this issue.

Readers should have a certain level of logical background: in other words, this monograph is not an introductory textbook for logical systems. After reading an introductory text, however, this book will be an excellent introduction and overview of non-standard logical and set systems of current interest.

Thus, this monograph is suited to researchers of the related fields of artificial intelligence as well as to Ph.D. course students who are motivated to find appropriate subjects of studies.

Overall, it should be emphasized again that the contents herein are unique and strongly stimulating future studies of the related areas of research, and the results of the authors should deeply be appreciated in view of their wide perspective and deep insight.

Tsukuba, Japan
September 2022

Sadaaki Miyamoto

Preface

Artificial Intelligence (AI) is the research area of science and engineering for intelligent machines, especially intelligent computer programs. One of the basic objectives is to study computer-based advanced knowledge representation and reasoning for developing knowledge-based systems.

It is very important to deal with common-sense reasoning in knowledge-based systems. If we employ a logic-based framework, classical logic is not suited for the purpose of describing common-sense reasoning. It is well known that there are several difficulties with logic-based approaches, e.g., the so-called Frame Problem.

This book explores a new approach to common-sense reasoning using epistemic situation calculus which integrates the ideas of situation calculus and epistemic logic. We try to formalize common-sense reasoning in the context of granular computing based on rough set theory.

The structure of the book is as follows. Chapter 1 gives the motivations and the organization of the present book. We first discuss the importance of the formalization of common-sense reasoning in knowledge-based systems. Second, we sketch our approach using epistemic situation calculus which integrates the ideas of situation calculus and epistemic logic. We try to formalize common-sense reasoning in the context of granular computing based on rough set theory. We also briefly show the organization of the book.

Chapter 2 introduces the foundations for *rough set theory* [1]. We outline Pawlak's motivating idea and give a technical exposition. Besides variable precision, rough set model is also presented with some applications based on rough set theory. In addition, we describe some non-classical logics. They are closely related to the foundations of rough set theory. We also provide the basics of modal and many-valued logic.

In Chap. 3, we present the consequence relation and Gentzen type sequent calculus for many-valued logics, and also describe the partial semantics interpreted with rough sets.

In Chap. 4, we present another deduction system based on tableaux calculus for four-valued logic and introduce the analytic tableaux as a basis for an automated deductive system.

In Chap. 5, we apply granular reasoning to the epistemic situation calculus ES of Lakemeyer and Levesque by interpreting actions as modalities and granules of possible worlds as states. The zoom reasoning proposed by Murai et al. is regarded as an epistemic action and is incorporated into the ES as an abstraction and refinement action by the granularity of the situation.

In Chap. 6, we discuss some aspects of the Frame Problem in the context of granular reasoning. We describe the essential concern behind the Frame Problem and discuss our point of view on the Frame Problem.

In Chap. 7, we give our conclusions about the book. First, we summarize the results, second, we show further technical questions which should be solved in our future work.

The book proposes a novel approach to common-sense reasoning. We assume that the reader has mastered the material ordinarily covered in AI and mathematical logic. We believe that the book is suitable for those, like experts and students, who wish to get involved in the field.

We are grateful to Prof. L. Jain, Prof. Y. Kudo, Prof. S. Miyamoto, and the referees for their constructive comments.

Kawasaki, Japan — Seiki Akama
Kōtō, Japan — Yotaro Nakayama
Chitose, Japan — Tetsuya Murai
September 2022

Contents

Chapter 1
Introduction

Abstract Chapter 1 gives the motivations and the organization of the present book. We first discuss the importance of the formalization of common-sense reasoning in knowledge-based systems. Second, we sketch our approach using epistemic situation calculus which integrates the ideas of situation calculus and epistemic logic. We try to formalize common-sense reasoning in the context of granular computing based on rough set theory. We also briefly show the organization of the book.

1.1 Motivations

Artificial Intelligence (AI) is the research area of science and engineering for intelligent machines, especially intelligent computer programs. One of the basic objectives is to study computer-based advanced knowledge representation and reasoning for developing knowledge-based systems.

It is very important to deal with *common-sense reasoning* in knowledge-based systems. If we employ a logic-based framework, *classical logic* is not suited for the purpose of describing common-sense reasoning. It is well known that there are several difficulties with logic-based approaches. e.g., the so-called *Frame Problem*.

This book explores a new approach to common-sense reasoning using epistemic situation calculus which integrates the ideas of situation calculus and epistemic logic. We try to formalize common-sense reasoning in the context of granular computing based on rough set theory. Thus, the proposed approach is based on *non-classical logic* rather than classical logic as a tool for reasoning and knowledge representation.

We generalize *epistemic situation calculus* (ES) of Lakemeyer and Levesque [2] which integrates ideas of situation calculus and epistemic logic for formalizing common-sense reasoning in the context of granular computing based on rough set theory.

Why do we need such a framework? For incomplete and inconsistent information in knowledge representation and reasoning, we assume that the concept of partiality

S. Akama et al., *Epistemic Situation Calculus Based on Granular Computing*,
Intelligent Systems Reference Library 239,
https://doi.org/10.1007/978-3-031-28551-6_1

and overcompleteness play an essential role. The partiality and overcompleteness of information are causes of incomplete or inconsistent epistemic states on the knowledge and lead to uncertain and ambiguous interpretations of the world.

Reasoning for partial and inconsistent information inherits ambiguity, and uncertainty induces incompleteness and inconsistency. Incompleteness and inconsistency are in fact caused by the lack of information and the excessiveness of information.

In classical logic, sentences are either true or false, and it is assumed that at any time, every sentence has exactly one of these two truth-values regardless of the available information about it. In knowledge representation, information is not complete. Therefore, we need *non-classical logic* as a model for reasoning with inconsistent and incomplete information.

We will treat here non-classical logic, also sometimes referred to as partial logic that is needed in order to formulate and understand the proposed framework of knowledge representation and reasoning. The idea of our knowledge representation and reasoning concern two main aspects of AI technology: cognitive adequacy and computational feasibility.

The goal is thus to develop an appropriate methodology for knowledge representation and reasoning systems. The point of departure is the kind of knowledge representation achieved by the epistemic situation calculus and reasoning system based on non-classical logic.

Non-classical logics can be classified into two classes. One class is a rival to classical logic, including intuitionistic logic and many-valued logic. The logics in this class deny some principles of classical logic. For example, many-valued logic denies the two-valuedness of classical logic, allowing several truth-values. Kleene's three-valued logic does not adopt the law of excluded middle $A \vee \neg A$ as an axiom. Consequently, many-valued logic can deal with the features of information like degree and vagueness.

The second class is an extension of classical logic. The logics in this class thus expand classical logic with some logical types of machinery. One of the most notable logics in the class is modal logic, which extends classical logic with modal operators to formalize the notions of necessity and possibility.

A lack of information and contradictory information is an epistemic state which is captured similar to ambiguity or uncertainty, but the gap and glut should be treated as distinguished. In the glut of information, the *undecidability* of the truth results from the lack of information, so the contradiction is also the problem of partiality of information.

Our position is that reasoning as a cognitive activity is based on the knowledge available to the cognitive agent through some form of representation. The goal of knowledge representation (KR) in computer science is to improve the existing KR technology and theory.

In the real world, an agent makes judgments and draws a conclusion even if he does not have complete information about a problem, and even if the information at hand contains inconsistency. This kind of practical reasoning is distinct from standard deductive and probabilistic reasoning, dealing with exact predicates without truth-value gaps excluding the possibility of incoherent information.

Thus, the aim is to establish a formal system that: (1) captures a practically relevant body of cognitive facilities employed by humans which can be reasonably implemented, and (2) extends the knowledge representation and reasoning capabilities of humans by capitalizing on the technical strength of the system including formal techniques to master gluts (situations where we are inclined to accept contradicting statements) and gaps (situations where we want to draw a conclusion but are uncertain because of lack of information) in a principled way.

In this book, we study how these fundamental theories–non-classical logics and their deduction systems–can work for the reasoning in the situation calculus that includes frame axioms used in the Frame Problem.

J. McCarthy and P. Hayes [3] first presented *the Frame Problem*. Moreover, Dennett [1] redefined it as the generalized Frame Problem. It is well known that the Frame Problem is the challenge of the representation of the effects of action in logic having to represent them in tractable fashion.

However, to many philosophers, the AI researchers' Frame Problem is suggestive of wider epistemological issues [1], and the Frame Problem gives epistemological issues to philosophers.

The problem is the possibility to limit the scope of the reasoning required to derive the consequences of an action. Moreover, more generally, the problem is that to account for an agent able to make decisions on the basis only of what is relevant to an ongoing situation without having explicitly to consider all that is not relevant.

The Frame Problem is a question of how to process only thc necessary information from the information making up the world and this is raised as the first serious problem in the world's description.

The second problem is the amount of processing of inference and is said as a problem due to processing with partiality. This can be seen as a problem of difficulty in identifying problems related to the subject of actions posed in the problem of Dennett's robot example. The fundamental problem on the descriptiveness of the Frame Problem is expected to be a problem of correspondence to partiality,

Our approach does not aim to give a perfect solution for the Frame Problem, but we provide the interpretation based on the partiality to the problem of description of the information, which is considered as the cause of it.

To consider the partiality of the information, we give another interpretation to the Frame Problem. The fundamental difficulty is here due to the impossibility of description for the world and situation. The Frame Problem was assumed to be based on a complete description of the required condition for actions.

On the other hand, granular reasoning is a reasoning framework based on the analytics method of information considering the partiality and approximation, and the degree of validity.

In this book, we study the knowledge representation and reasoning for an agent in the world in which partiality plays a key concept. In other words, we have to cope with knowledge representation with incomplete and inconsistent information adequately.

Therefore we adopt many-valued logics and modal logics, which are non-classical logics, and rough set theory, which can serve as the theoretical basis for data analysis and the semantic basis for our non-classical logics.

When we focus on the knowledge and the recognition of an agent, we need to handle partiality, ambiguity, vagueness, incompleteness and inconsistency. This phenomenon cannot be captured by classical bivalence logic and we need plausible and defeasible system as underlying theory.

Partiality is the primary concern in the research of artificial intelligence, and it is still considered as a significant problem in this field. To handle the partiality for knowledge representation, there are various approaches.

In this research, as a tool, we focus on the rough set theory, granular computing, many-valued logics, and modal logics. In fact, these theories and logics are closely related.

1.2 Organization

The contents of the book are essentially based on those presented in Nakayama [4]. This book consists of two parts: The first is about the deduction system with many-valued logics and the second is about the granular reasoning in the epistemic situation calculus.

The deduction system treats non-classical logic as a basis for reasoning with partial information, and granular reasoning is embedded in the ES for describing common-sense reasoning including non-monotonic reasoning with partial semantics.

The structure of the book is as follows. In Chap. 2, we introduce the foundations for *rough set theory* [5]. We outline Pawlak's motivating idea and give a technical exposition. Besides variable precision rough set model is also presented with some applications based on rough set theory. In addition, we describe some non-classical logics. They are closely related to the foundations of rough set theory. We also provide the basics of modal and many-valued logic.

In Chap. 3, we present the consequence relation and Gentzen type sequent calculus for many-valued logics, and also describe the partial semantics interpreted with rough sets.

In Chap. 4, we present another deduction system based on tableaux calculus for four-valued logic and introduce the analytic tableau as a basis for an automated deductive system.

In Chap. 5, we apply granular reasoning to the epistemic situation calculus ES of Lakemeyer and Levesque by interpreting actions as modalities and granules of possible worlds as states. The zoom reasoning proposed by Murai et al. is regarded as an epistemic action and is incorporated into the ES as an abstraction and refinement action by the granularity of the situation.

In Chap. 6, we discuss some aspects of the Frame Problem in the context of granular reasoning. We describe the essential concern behind the Frame Problem and discuss our point of view for the Frame Problem.

In Chap. 7, we give our conclusions of the book. First, we summarize the results, Second, we show further technical questions which should be solved in our future work.

References

1. Dennett, D.: Cognitive wheels: The frame problem of AI. In: Hookway, C. (ed.) Minds, Machines and Evolution, pp. 129–151. Cambridge University Press, Cambridge (1984)
2. Lakemeyer, G., Levesque, H.: A semantic characterization of a useful fragment of the situation calculus with knowledge. Artif. Intell. **175**, 142-164 (2011)
3. McCarthy, J., Hayes, P.: Some philosophical problems from the standpoint of artificial intelligence. Mach. Intell. **4**, 463-502 (1969)
4. Nakayama, Y.: A study on the epistemic situation calculus in the framework of granular reasoning. Ph.D. Thesis, Chitose Institute of Science and Technology (CIST), Chitose, Hokkaido, Japan (2020)
5. Pawlak, Z.: Rough sets. Int. J. Comput. Inf. Sci. **11**, 341-356 (1982)

Chapter 2
Preliminaries

Abstract Chapter 2 introduces the foundations for *rough set theory*. We outline Pawlak's motivating idea and give a technical exposition. Besides variable precision rough set model is also presented with some applications based on rough set theory. In addition, we describe some non-classical logics. They are closely related to the foundations of rough set theory. We also provide the basics of modal and many-valued logic.

2.1 Rough Set Theory

This section describes the foundations for rough set theory. We outline Pawlak's motivating idea and give a technical exposition. Basics of Pawlak's rough set theory and variable precision rough set model are presented with some related topics.

Rough set theory is interesting theoretically as well as practically, and a quick survey on the subject, including overview, history and applications, is helpful to the readers.

2.1.1 Pawlak's Rough Set Theory

Rough set theory, proposed by Pawlak [25], provides a theoretical basis of sets based on approximation concepts. A rough set can be seen as an approximation of a set. These approximations can be described by two operators on subsets of the universe.

Rough set theory is, in particular, helpful in extracting knowledge from data tables and it has been successfully applied to the areas such as data analysis, decision making, machine learning and other various applications

We begin with an exposition of Pawlak's approach to rough set theory based on Pawlak [26]. His motivation is to provide a theory of knowledge and classification by introducing a new concept of set, i.e., rough set.

By *object*, we mean anything we can think of, for example, real things, states, abstract concepts, etc. We can assume that knowledge is based on the ability to

S. Akama et al., *Epistemic Situation Calculus Based on Granular Computing*,
Intelligent Systems Reference Library 239,
https://doi.org/10.1007/978-3-031-28551-6_2

classify objects. Thus, knowledge is necessarily connected with the variety of classification patterns related to specific parts of the real or abstract world, called the *universe of discourse* (or the *universe*).

Now, we turn to a formal presentation. We assume the usual notation for set theory. Let U be non-empty finite set of objects called the universe of discourse. Any subset $X \subseteq U$ of the universe is called a *concept* or a *category* in U. Any family of concepts in U is called *knowledge* about U. Note that the empty set $\emptyset$ is also a concept.

We mainly deal with concepts which form a partition (classification) of a certain universe U, i.e. in families $C = \{X_1, X_2, \dots, X_n\}$ such that $X_i \subseteq U$, $X_i \neq \emptyset$, $X_i \cap X_j \neq \emptyset$ for $i \neq j, i, j = 1, \dots, n$ and $\bigcup X_i = U$. A family of classifications over U is called a knowledge base over U.

Classifications can be specified by using *equivalence relations*. If R is an equivalence relation over U, then U/R means the family of all equivalence classes of R (or classification of U) referred to as categories or concepts of R. $[x]_R$ denotes a category in R containing an element $x \in U$.

A knowledge base is defined as a relational system, $K = (U, \mathbf{R})$, where $U \neq \emptyset$ is a finite set called the *universe*, and $\mathbf{R}$ is a family of equivalence relations over U. $IND(K)$ means the family of all equivalence relations defined in K, i.e., $IND(K) = \{IND(\mathbf{P}) | \emptyset \neq \mathbf{P} \subseteq \mathbf{R}\}$. Thus, $IND(K)$ is the minimal set of equivalence relations, containing all elementary relations of K, and closed under set-theoretical intersection of equivalence relations.

If $\mathbf{P} \subseteq \mathbf{R}$ and $\mathbf{P} \neq \emptyset$, then $\cap \mathbf{P}$ denotes the intersection of all equivalence relations belonging to $\mathbf{P}$, denoted $IND(\mathbf{P})$, called an indiscernibility relation of $\mathbf{P}$. It is also an equivalence relation, and satisfies:

$$[x]_{IND(P)} = \bigcap_{R \in \mathbf{P}} [x]_R.$$

Thus, the family of all equivalence classes of the equivalence relation $IND(\mathbf{P})$, i.e., $U/IND(\mathbf{P})$ denotes knowledge associated with the family of equivalence relations $\mathbf{P}$. For simplicity, we will write $U/\mathbf{P}$ instead of $U/IND(\mathbf{P})$.

$\mathbf{P}$ is also called $\mathbf{P}$-*basic* knowledge. Equivalence classes of $IND(\mathbf{P})$ are called basic categories (concepts) of knowledge $\mathbf{P}$. In particular, if $Q \in \mathbf{R}$, then Q is called a *Q-elementary* knowledge (about U in K) and equivalence classes of Q are referred to as *Q-elementary concepts (categories)* of knowledge $\mathbf{R}$.

Now, we describe the fundamentals of rough sets. Let $X \subseteq U$ and $\mathbf{R}$ be an equivalence relation. We say that X is *R-definable* if X is the union of some *R-basic categories*; otherwise X is *R-undefinable*.

The R-definable sets are those subsets of the universe which can be exactly defined in the knowledge base K, whereas the R-undefinable sets cannot be defined in K. The R-definable sets are called *R-exact* sets, and R-undefinable sets are called *R-inexact* or *R-rough*.

Set $X \subseteq U$ is called *exact* in K if there exists an equivalence relation $R \in IND(K)$ such that X is *R-exact*, and X is said to be rough in K if X is *R-rough* for any $R \in IND(K)$.

Observe that rough sets can also be defined approximately by using two exact sets, referred as a lower and an upper approximation of the set. Now, rough set theory is outlined.

Definition 2.1 A knowledge base K is a pair $K = (U, R)$, where U is a universe of objects, and $\mathbf{R}$ is a set of equivalence relations on the objects in U.

Definition 2.2 Let $R \in \mathbf{R}$ be an equivalence relation of the knowledge base $K = (U, R)$ and X any subset of U. Then, the lower and upper approximations of X for R are defined as follows:

$\underline{R}X = \bigcup\{Y \in U/R \mid Y \subseteq X\}$,

$\overline{R}X = \bigcup\{Y \in U/R \mid Y \cap X \neq 0\}$.

$\underline{R}X$ is called the R-lower approximation and $\overline{R}X$ the R-upper approximation of X, respectively. They will be simply called the *lower-approximation* and the *upper-approximation* if the context is clear.

It is also possible to define the lower approximation and the upper approximation in the following two equivalent forms:

$\underline{R}X = \{x \in U \mid [x]_{\mathrm{R}} \subseteq X\}$,

$\overline{R}X = \{x \in U \mid [x]_{\mathrm{R}} \cap X \neq \emptyset\}$.

or

$x \in \underline{R}X$ iff $[x]_R \subseteq X$,

$x \in \underline{R}X$ iff $[x]_R \cap X \neq \emptyset$.

The above three are interpreted as follows. The set $\underline{R}X$ is the set of all elements of U which can be certainly classified as elements of X in the knowledge R. The set $\overline{R}X$ is the set of elements of U which can be possibly classified as elements of X in R.

We define *R-positive region* ($POS_R(X)$), *R-negative region* ($NEG_R(X)$), and *R-borderline region* ($BN_R(X)$) of X as follows:

Definition 2.3 If $K = (U, R)$, $R \in R$, and $X \subseteq U$, then the R-positive, R-negative, and R-boundary regions of X with respect to R are defined respectively as follows:

$$POS_R(X) = \underline{R}X,$$
$$NEG_R(X) = U - \overline{R}X,$$
$$BN_R(X) = \overline{R}X - \underline{R}X.$$

The positive region $POS_R(X)$ (or the lower approximation) of X is the collection of those objects which can be classified with full certainty as members of the set X, using knowledge R.

The negative region $NEG_R(X)$ is the collection of objects with which it can be determined without any ambiguity, employing knowledge R, that they do not belong to the set X, that is, they belong to the complement of X.

The borderline region $BN_R(X)$ is the set of elements which cannot be classified either to X or to $-X$ in R. It is the undecidable area of the universe, i.e., none of the objects belonging to the boundary can be classified with certainty into X or $-X$ as far as R is concerned.

Now, we list basic formal results. Their proofs may be found in Pawlak [26]. Proposition 2.1 is obvious.

Proposition 2.1 *The following hold:*

1. *X is R-definable iff $\underline{R}X = \overline{R}X$*
2. *X is rough with respect to $\underline{R}X \neq \overline{R}X$*

Proposition 2.2 shows the basic properties of approximations:

Proposition 2.2 *The R-lower and R-upper approximations satisfy the following properties:*

1. $\underline{R}X \subseteq X \subseteq \overline{R}X$
2. $\underline{R}\emptyset = \overline{R}\emptyset = \emptyset, \underline{R}U = \overline{R}U = U$
3. $\overline{R}(X \cup Y) = \overline{R}X \cup \overline{R}Y$
4. $\underline{R}(X \cap Y) = \underline{R}X \cap \underline{R}Y$
5. $X \subseteq Y$ *implies* $\underline{R}X \subseteq \underline{R}Y$
6. $X \subseteq Y$ *implies* $\overline{R}X \subseteq \overline{R}Y$
7. $\underline{R}(X \cup Y) \subseteq \underline{R}X \cup \underline{R}Y$
8. $\overline{R}(X \cap Y) \subseteq \overline{R}X \cap \overline{R}Y$
9. $\underline{R}(-X) = -\overline{R}X$
10. $\overline{R}(-X) = -\underline{R}X$
11. $\underline{R}\underline{R}X = \overline{R}\underline{R}X = \underline{R}X$
12. $\overline{R}\overline{R}X = \underline{R}\overline{R}X = \overline{R}X$

The concept of approximations of sets can also be applied to that of *membership relation*. In rough set theory, since the definition of a set is associated with knowledge about the set, a membership relation must be related to the knowledge.

Then, we can define two membership relations $\underline{\in}_R$ and $\overline{\in}_R$. $x \underline{\in}_R$ reads "x surely belongs to X" and $x \overline{\in}_R$ reads "x possibly belongs to X". $\underline{\in}_R$ and $\overline{\in}_R$ are called the *R-lower membership* and *R-upper membership*, respectively.

Proposition 2.3 states the basic properties of membership relations:

Proposition 2.3 *The R-lower and R-upper membership relations satisfy the following properties:*

1. $x \underline{\in}_R X$ *implies* $x \in X$ *implies* $x \overline{\in}_R$
2. $X \subseteq Y$ *implies* ($x \underline{\in}_R X$ *implies* $x \underline{\in}_R Y$ *and* $x \overline{\in}_R X$ *implies* $x \overline{\in}_R Y$)
3. $x \overline{\in}_R (X \cup Y)$ *iff* $x \overline{\in}_R X$ *or* $x \overline{\in}_R Y$
4. $x \underline{\in}_R (X \cap Y)$ *iff* $x \underline{\in}_R X$ *and* $x \underline{\in}_R Y$
5. $x \underline{\in}_R X$ *or* $x \underline{\in}_R Y$ *implies* $x \underline{\in}_R (X \cup Y)$
6. $x \overline{\in}_R (X \cap Y)$ *implies* $x \overline{\in}_R X$ *and* $x \overline{\in}_R Y$
7. $x \underline{\in}_R (-X)$ *iff non* $x \overline{\in}_R X$
8. $x \overline{\in}_R (-X)$ *iff non* $x \underline{\in}_R X$

Approximate (rough) equality is the concept of equality in rough set theory. Three kinds of *approximate equality* can be introduced. Let $K = (U, \mathbf{R})$ be a knowledge base, $X, Y \subseteq U$ and $R \in IND(K)$.

1. Sets X and Y are bottom R-equal $(X \overline{\sim}_R Y)$ if $\underline{R}X = \underline{R}Y$
2. Sets X and Y are top R-equal $(X \simeq_R Y)$ if $\overline{R}X = \overline{R}Y$
3. Sets X and Y are R-equal $(X \approx_R Y)$ if $(X \overline{\sim}_R Y)$ and $(X \simeq_R Y)$

These equalities are equivalence relations for any indiscernibility relation R. They are interpreted as follows: $(X \overline{\sim}_R Y)$ means that positive example of the sets X and Y are the same, $(X \simeq_R Y)$ means that negative example of the sets X and Y are the same, and $(X \approx_R Y)$ means that both positive and negative examples of X and Y are the same.

These equalities satisfy the following proposition (we omit subscript R for simplicity):

Proposition 2.4 *For any equivalence relation, we have the following properties:*

1. *(1)* $X \overline{\sim} Y$ *iff* $X \cap X \overline{\sim} Y$ *and* $X \cap Y \overline{\sim} Y$
2. $X \simeq Y$ *iff* $X \cup Y \simeq X$ *and* $X \cup Y \simeq Y$
3. *If* $X \simeq X'$ *and* $Y \simeq Y'$, *then* $X \cup Y \simeq X' \cup Y'$
4. *If* $X \overline{\sim} X'$ *and* $Y \overline{\sim} Y'$, *then* $X \cap Y \overline{\sim} X' \cap Y'$
5. *If* $X \simeq X'Y$, *then* $X \cup -Y \simeq U$
6. *If* $X \overline{\sim} Y$, *then* $X \cap -Y \overline{\sim} \emptyset$
7. *If* $X \subseteq Y$ *and* $Y \simeq \emptyset$, *then* $X \simeq \emptyset$
8. *If* $X \subseteq Y$ *and* $Y \simeq U$, *then* $X \simeq U$
9. $X \simeq Y$ *iff* $-X \overline{\sim} -Y$
10. *If* $X \overline{\sim} \emptyset$ *or* $Y \overline{\sim} \emptyset$, *then* $X \cap Y \overline{\sim} \emptyset$
11. *If* $X \simeq U$ *or* $Y \simeq U$, *then* $X \cup Y \simeq U$.

The following Proposition 2.5 shows that lower and upper approximations of sets can be expressed by rough equalities:

Proposition 2.5 *For any equivalence relation R:*

1. $\underline{R}X$ *is the intersection of all* $Y \subseteq U$ *such that* $X \overline{\sim}_R Y$
2. $\overline{R}X$ *is the union of all* $Y \subseteq U$ *such that* $X \simeq_R Y$.

Similarly, we can define rough inclusion of sets. It is possible to define three kinds of rough inclusions.

Let $X = (U, \mathbf{R})$ be a knowledge base, $X, Y \subseteq U$, and $R \in IND(K)$. Then, we have:

1. Set X is bottom R-included in Y $(X \overset{\subset}{\sim}_R Y)$ iff $\underline{R}X \subseteq \underline{R}Y$
2. Set X is top R-included in Y $(X \tilde{\subset}_R Y)$ iff $\overline{R}X \subseteq \overline{R}Y$
3. Set X is R-included in Y $(X \overset{\sim}{\underset{\sim}{\subset}}_R Y)$ iff $(X \overset{\subset}{\sim}_R Y)$ and $(X \tilde{\subset}_R Y)$

Proposition 2.6 *Rough inclusion satisfies the following:*

(1) If $X \subseteq Y$, then $X \underset{\sim}{\subset} Y$, $X \overset{\sim}{\subset} Y$ and $X \overset{\sim}{\underset{\sim}{\subset}} Y$

(2) If $X \underset{\sim}{\subset} Y$ and $Y \underset{\sim}{\subset} X$, then $X \overline{\sim} Y$

(3) If $X \overset{\sim}{\subset} Y$ and $Y \overset{\sim}{\subset} X$, then $X \simeq Y$

(4) If $X \overset{\sim}{\underset{\sim}{\subset}} Y$ and $Y \overset{\sim}{\underset{\sim}{\subset}} X$, then $X \approx Y$

(5) If $X \overset{\sim}{\subset} Y$ iff $X \cup Y \simeq Y$

(6) If $X \underset{\sim}{\subset} Y$ iff $X \cap Y \overline{\sim} Y$

(7) If $X \subseteq Y$, $X \overline{\sim} X'$ and $Y \overline{\sim} Y'$, then $X' \underset{\sim}{\subset} Y'$

(8) If $X \subseteq Y$, $X \simeq X'$ and $Y \simeq Y'$, then $X' \overset{\sim}{\subset} Y'$

(9) If $X \subseteq Y$, $X \approx X'$ and $Y \approx Y'$, then $X' \overset{\sim}{\underset{\sim}{\subset}} Y'$

(10) If $X' \overset{\sim}{\subset} X$ and $Y' \overset{\sim}{\subset} Y$, then $X' \cup Y' \overset{\sim}{\subset} X \cup Y$

(11) $X' \underset{\sim}{\subset} X$ and $Y' \underset{\sim}{\subset} Y$, then $X' \cap Y' \underset{\sim}{\subset} X \cap Y$

(12) $X \cap Y \underset{\sim}{\subset} Y \overset{\sim}{\subset} X \cup Y$

(13) If $X \underset{\sim}{\subset} Y$ and $X \overline{\sim} Z$, then $Z \underset{\sim}{\subset} Y$

(14) If $X \overset{\sim}{\subset} Y$ and $X \simeq Z$, then $Z \overset{\sim}{\subset} Y$

(15) If $X \overset{\sim}{\underset{\sim}{\subset}} Y$ and $X \approx Z$, then $Z \overset{\sim}{\underset{\sim}{\subset}} Y$

The above properties are not valid if we replace $\overline{\sim}$ by $\simeq$ (or conversely). If R is an equivalence relation, then all three inclusions reduce to ordinary inclusion.

Note that $\underset{\sim}{\subset}_R$, $\overset{\sim}{\subset}_R$ and $\overset{\sim}{\underset{\sim}{\subset}}_R$ are quasi ordering relations. They are called the lower, upper and rough inclusion relation, respectively. Observe that rough inclusion of sets does not imply the inclusion of sets.

The following proposition shows the properties of rough inclusion:

Proposition 2.7 *Rough inclusion satisfies the following:*

(1) If $X \subseteq Y$, then $X \underset{\sim}{\subset} Y$, $X \overset{\sim}{\subset} Y$ and $X \overset{\sim}{\underset{\sim}{\subset}} Y$

(2) If $X \underset{\sim}{\subset} Y$ and $Y \underset{\sim}{\subset} X$, then $X \overline{\sim} Y$

(3) If $X \overset{\sim}{\subset} Y$ and $Y \overset{\sim}{\subset} X$, then $X \simeq Y$

(4) If $X \overset{\sim}{\underset{\sim}{\subset}} Y$ and $Y \overset{\sim}{\underset{\sim}{\subset}} X$, then $X \approx Y$

(5) If $X \overset{\sim}{\subset} Y$ iff $X \cup Y \simeq Y$

(6) If $X \underset{\sim}{\subset} Y$ iff $X \cap Y \overline{\sim} Y$

(7) *If* $X \subseteq Y$, $X \eqsim X'$ *and* $Y \eqsim Y'$, *then* $X' \underset{\sim}{\subset} Y'$

(8) *If* $X \subseteq Y$, $X \simeq X'$ *and* $Y \simeq Y'$, *then* $X' \overset{\sim}{\subset} Y'$

(9) *If* $X \subseteq Y$, $X \approx X'$ *and* $Y \approx Y'$, *then* $X' \underset{\sim}{\overset{\sim}{\subset}} Y'$

(10) *If* $X' \overset{\sim}{\subset} X$ *and* $Y' \overset{\sim}{\subset} Y$, *then* $X' \cup Y' \overset{\sim}{\subset} X \cup Y$

(11) $X' \underset{\sim}{\subset} X$ *and* $Y' \underset{\sim}{\subset} Y$, *then* $X' \cap Y' \underset{\sim}{\subset} X \cap Y$

(12) $X \cap Y \underset{\sim}{\subset} Y \overset{\sim}{\subset} X \cup Y$

(13) *If* $X \underset{\sim}{\subset} Y$ *and* $X \eqsim Z$, *then* $Z \underset{\sim}{\subset} Y$

(14) *If* $X \overset{\sim}{\subset} Y$ *and* $X \simeq Z$, *then* $Z \overset{\sim}{\subset} Y$

(15) *If* $X \underset{\sim}{\overset{\sim}{\subset}} Y$ *and* $X \approx Z$, *then* $Z \underset{\sim}{\overset{\sim}{\subset}} Y$

The above properties are not valid if we replace $\eqsim$ by $\simeq$ (or conversely). If R is an equivalence relation, then all three inclusions reduce to ordinary inclusion.

2.1.2 Variable Precision Rough Set

The *variable precision rough set model* (VPRS model) proposed by Ziarko [34] is an extension of Pawlak's rough set theory, which provides a theoretical basis to treat probabilistic or inconsistent information in the framework of rough sets.

Ziarko generalized Pawlak's original rough set model in Ziarko [34], which is called the *variable precision rough set model* (VPRS model) to overcome the inability to model uncertain information, and is directly derived from the original model without any additional assumptions.

In addition, VPRS model is an extension of Pawlak's rough set theory, which provides a theoretical basis to treat probabilistic or inconsistent information in the framework of rough sets.

As the limitations of Pawlak's rough set model, Ziarko discussed two points. First, it cannot provide a classification with a controlled degree of uncertainty. Second, some level of uncertainty in the classification process gives a deeper or better understanding of data analysis.

VPRS model generalizes the standard set inclusion relation, capable of allowing for some degree of misclassification in the largely correct classification.

Let X and Y be non-empty subsets of a finite universe U. X is included in Y, denoted $Y \supseteq X$, if for all $e \in X$ implies $e \in Y$. Here, we introduce the measure $c(X, Y)$ of the relative degree of misclassification of the set X with respect to set Y defined as:

$$c(X, Y) =_{\text{def}} \begin{cases} 1 - \dfrac{|X \cap Y|}{|X|}, & \text{if} X \neq \emptyset, \\ 0, & \text{otherwise.} \end{cases},$$

where $|X|$ represents the cardinality of the set X.

VPRS is based on the majority inclusion relation. Let $X, Y \subseteq U$ be any subsets of U. The majority inclusion relation is defined by the following measure $c(X, Y)$ of the relative degree of misclassification of X with respect to Y.

The quantity $c(X, Y)$ will be referred to as the relative classification error. The actual number of misclassification is given by the product $c(X, Y) * card(X)$ which is referred to as an absolute classification error.

We can define the inclusion relationship between X and Y without explicitly using a general quantifier:

$X \subseteq Y$ iff $c(X, Y) = 0$.

The majority requirement implies that more than 50% of X elements should be in common with Y. The specified majority requirement imposes an additional requirement. The number of elements of X in common with Y should be above 50% and not below a certain limit, e.g. 85%.

Formally, the majority inclusion relation $\overset{\beta}{\subseteq}$ with a fixed precision $\beta \in [0, 0.5)$ is defined using the relative degree of misclassification as follows:

$X \overset{\beta}{\subseteq} Y$ iff $c(X, Y) \leq \beta$,

where precision β provides the limit of permissible misclassification.

The above definition covers the whole family of β-majority relation. However, the majority inclusion relation does not have the transitivity relation.

The following two propositions indicate some useful properties of the majority inclusion relation:

Proposition 2.8 *If $A \cap B = \emptyset$ and $B \overset{\beta}{\supseteq} X$, then it is not true that $A \overset{\beta}{\supseteq} X$.*

Proposition 2.9 *If $\beta_1 < \beta_2$, then $Y \overset{\beta_2}{\supseteq} X$ implies $Y \overset{\beta_2}{\supseteq} X$.*

For the VPRS-model, we define the approximation space as a pair $A = (U, R)$, where U is a non-empty finite universe and R is the equivalence relation on U. The equivalence relation R, referred to as an indiscernibility relation, corresponds to a partitioning of the universe U into a collection of equivalence classes or elementary set $R' = \{E_1, E_2, \ldots, E_n\}$.

Using a majority inclusion relation instead of the inclusion relation, we can obtain generalized notions of β-lower approximation (or β-positive region $POSR_\beta(X)$) of the set $U \supseteq X$:

$$\underline{R}_\beta X = \bigcup\{E \in R^* : X \overset{\beta}{\supseteq} E\} \text{ or, equivalently,}$$
$$\underline{R}_\beta X = \bigcup\{E \in R^* : c(E, X) \leq \beta\}$$

The β-upper approximation of the set $U \supseteq X$ can also be defined as follows:

$$\overline{R}_\beta X = \bigcup\{E \in R^* : c(E, X) < 1 - \beta\}$$

The β-boundary region of a set is given by

$$BNR_\beta X = \bigcup\{E \in R^* : \beta < c(E, X) < 1 - \beta\}.$$

The β-negative region of X is defined as a complement of the β-upper approximation:

$$NEGR_\beta X = \bigcup\{E \in R^* : c(E, X) \geq 1 - \beta\}.$$

The lower approximation of the set X can be interpreted as the collection of all those elements of U which can be classified into X with the classification error not greater than β.

The β-negative region of X is the collection of all those elements of U which can be classified into the complement of X, with the classification error not greater than β. The latter interpretation follows from Proposition 2.10:

Proposition 2.10 *For every $X \subseteq Y$, the following relationship is satisfied:*

$$POSR_\beta(-X) = NEGR_\beta X.$$

The β-boundary region of X consists of all those elements of U which cannot be classified either into X or into $-X$ with the classification error not greater than β.

Notice here that the law of excluded middle, i.e., $p \vee \neg p$, where $\neg p$ is the negation of p, holds in general for imprecisely specified sets.

Finally, the β-upper approximation $\overline{R}_\beta X$ of X includes all those elements of U which cannot be classified into $-X$ with the error not greater than β. If $\beta = 0$ then the original rough set model is a special case of VPRS-model, as the following proposition states:

Proposition 2.11 *Let X be an arbitrary subset of the universe U:*

1. *$\underline{R}_0 X = \underline{R}X$, where $\underline{R}X$ is a lower approximation defined as $\underline{R}X = \bigcup\{E \in R^* : X \supseteq E\}$*
2. *$\overline{R}_0 X = \overline{R}X$, where $\overline{R}X$ is an upper approximation defined as $\overline{R}X = \bigcup\{E \in R^* : E \cap X \neq \emptyset\}$*
3. *$BNR_0 X = BN_R X$, where $BN_R X$ is the set X boundary region defined as $BN_R X = \overline{R}X - \underline{R}X$*
4. *$NEGR_0 X = NEG_R X$, where $NEG_R X$ is the set X negative region defined as $NEG_R X = U - \overline{R}X$*

Proposition 2.12 *If $0 \leq \beta < 0.5$ then the properties listed in Proposition 2.10 and the following are also satisfied:*

$\underline{R}_\beta X \supseteq \underline{R}X,$
$\overline{R}X \supseteq \overline{R}_\beta X,$
$BN_R X \supseteq BNR_\beta X,$
$NEGR_\beta X \supseteq NEG_R X.$

Intuitively, with the decrease of the classification error β the size of the positive and negative regions of X will shrink, whereas the size of the boundary region will grow.

With the reduction of β fewer elementary sets will satisfy the criterion for inclusion in β-positive or β-negative regions. Thus, the size of the boundary will increase. The reverse process can be done with the increase of β.

Proposition 2.13 *With the β approaching the limit* 0.5, *i.e.*, $\beta \to 0.5$, *we obtain the following:*

$\underline{R}_\beta X \to \underline{R}_{0.5} X = \bigcup\{E \in R^* : c(E, X) < 0.5\},$
$\overline{R}_\beta X \to \overline{R}_{0.5} X = \bigcup\{E \in R^* : c(E, X) \le 0.5\},$
$BNR_\beta X \to BNR_{0.5} X = \bigcup\{E \in R^* : c(E, X) = 0.5\},$
$NEGR_\beta X \to NEGR_{0.5} X = \bigcup\{E \in R^* : c(E, X) > 0.5\}.$

The set $BNR_{0.5}X$ is called an absolute boundary of X because it is included in every other boundary region of X.

The following Proposition 2.14 summarizes the primary relationships between set X discernibility regions computed on 0.5 accuracy level and higher levels.

Proposition 2.14 *For boundary regions of X, the following hold:*

$BRN_{0.5}X = \bigcap_\beta BNR_\beta X,$

$R0.5X = \bigcap_\beta \overline{R}_\beta X,$

$R0.5X = \bigcap_\beta \underline{R}_\beta X,$

$NEGR0.5X = \bigcap_\beta NEGR_\beta X.$

The absolute boundary is very "narrow", consisting only of those sets which have 50/50 aplite of elements among set X interior and its exterior. All other elementary sets are classified either into positive region $\underline{R}_{0.5}X$ or the negative region $NEGR_{0.5}$.

We turn to the measure of approximation. To express the degree with which a set X can be approximately characterized by means of elementary sets of the approximation space $A = (U, R)$, we will generalize the accuracy measure introduced in Pawlak [25].

The β-accuracy for $0 \le \beta < 0.5$ is defined as

$$\alpha(R, \beta, X) = card(\underline{R}_\beta X)/card(\overline{R}_\beta X).$$

The β-accuracy represents the imprecision of the approximate characterization of the set X relative to assumed classification error β.

Note that with the increase of β the cardinality of the β-upper approximation will tend downward and the size of the β-lower approximation will tend upward which leads to the conclusion that is consistent with intuition that relative accuracy may increase at the expense of a higher classification error.

The notion of discernibility of set boundaries is relative. If a large classification error is allowed then the set X can be highly discernible within assumed classification limits. When smaller values of the classification tolerance are assumed it may become more difficult to discern positive and negative regions of the set to meet the narrow tolerance limits.

The set X is said to be β-discernible if its β-boundary region is empty or, equivalently, if

$$\overline{R_{\beta}}X = \overline{R}_{\beta}X.$$

For the β-discernible sets the relative accuracy $\alpha(R, \beta, X)$ is equal to unity. The discernible status of a set changes depending on the value of β. In general, the following properties hold:

Proposition 2.15 *If X is discernible on the classification error level* $0 \leq \beta < 0.5$, *then X is also discernible at any level* $\beta_1 > \beta$.

Proposition 2.16 *If* $\overline{R}_{0.5}X \neq \underline{R}_{0.5}X$, *then X is not discernible on every classification error level* $0 \leq \beta < 0.5$.

Proposition 2.16 emphasizes that a set with a non-empty absolute boundary can never be discerned. In general, one can easily demonstrate the following:

Proposition 2.17 *If X is not discernible on the classification error level* $0 \leq \beta < 0.5$, *then X is also not discernible at any level* $\beta_1 < \beta$.

Any set X which is not discernible for every β is called indiscernible or absolutely rough. The set X is absolutely rough iff $BNR_{0.5}X \neq \emptyset$. Any set which is not absolutely rough will be referred to as relatively rough or weakly discernible.

For each relatively rough set X, there exists such a classification error level β that X is discernible on this level.

Let $NDIS(R, X) = \{0 \leq \beta < 0.5 : BNR_{\beta}(X) \neq \emptyset\}$. Then, $NDIS(R, X)$ is a range of all those β values for which X is indiscernible.

The least value of classification error β which makes X discernible will be referred to as discernibility threshold. The value of the threshold is equal to the least upper bound $\zeta(R, X)$ of $NDIS(X)$, i.e.,

$$\zeta(R, X) = sup\ NDIS(R, X).$$

Proposition 2.18 states a simple property which can be used to find the discernibility threshold of a weakly discernible set X:

Proposition 2.18 $\zeta(R, X) = max(m1, m2)$, *where*

$$m_1 = 1 - min\{c(E, X) : E \in R^* \; and \; 0.5 < c(E, X)\},$$
$$m2 = max\{c(E, X) : E \in R^* \; and \; c(E, X) < 0.5\}.$$

The discernibility threshold of the set X equals a minimal classification error β which can be allowed to make this set β-discernible. We give some fundamental properties of β-approximations.

Proposition 2.19 *For every* $0 \leq \beta < 0.5$, *the following hold:*

(1a) $X \supseteq \underline{R}_{\beta} X$
(1b) $\overline{R}_{\beta} X \supseteq \underline{R}_{\beta} X$
(2) $\underline{R}_{\beta} \emptyset = \overline{R}_{\beta} \emptyset = \emptyset; \underline{R}_{\beta} U = \overline{R}_{\beta} U = U$
(3) $\overline{R}_{\beta}(X \cup Y) \supseteq \overline{R}_{\beta} X \cup \overline{R}_{\beta} Y$
(4) $\underline{R}_{\beta} X \cap \underline{R}_{\beta} Y \supseteq \underline{R}_{\beta}(X \cap Y)$
(5) $\underline{R}_{\beta}(X \cup Y) \supseteq \underline{R}_{\beta} \cup \underline{R}_{\beta} Y$
(6) $\overline{R}_{\beta} X \cap \overline{R}_{\beta} Y \supseteq \overline{R}_{\beta}(X \cap Y)$
(7) $\underline{R}_{\beta}(-X) = -\overline{R}_{\beta}(X)$
(8) $\overline{R}_{\beta}(-X) = -\underline{R}_{\beta}(X)$

We finish the outline of variable precision rough set model, which can be regarded as a direct generalization of the original rough set model. Consult Ziarko [34] for more details. As we will be discussed later, it plays an important role in our approach to rough set based reasoning.

Shen and Wang [32] proposed the VPRS model over two universes using inclusion degree. They introduced the concepts of the reverse lower and upper approximation operators and studied their properties. They introduced the approximation operators with two parameters as a generalization of the VPRS-model over two universes.

2.1.3 Decision Logic

Pawlak developed *decision logic* (DL) for reasoning about knowledge. His main goal is reasoning about knowledge concerning reality. Knowledge is represented as a value-attribute table, called *knowledge representation system*.

There are several advantages to represent knowledge in tabular form. The data table can be interpreted differently, namely it can be formalized as a logical system. The idea leads to decision logic.

We review the foundations of rough set-based decision logic [25, 26]. Let $S = (U, A)$ be a knowledge representation system. The language of DL consists of atomic formulas, which are attribute-value pairs, combined with logical connectives to form compound formulas. The alphabet of the language consists of:

1. A: the set of attribute constants
2. $V = \bigcup V_a$: the set of attribute constants $a \in A$

3. Set $\{\sim, \vee, \wedge, \rightarrow, \equiv\}$ of propositional connectives, called negation, disjunction, conjunction, implication and equivalence, respectively.

The set of formulas in *DL*-language is the least set satisfying the following conditions:

1. Expressions of the form (a, v), or in short a_v, called *atomic formulas*, are formulas of DL-language for any $a \in A$ and $v \in V_a$.
2. If ϕ and ψ are formulas of *DL*-language, then so are $\sim, \phi$, $(\phi \vee \psi)$, $(\phi \wedge \psi)$, $(\phi \rightarrow \psi)$ and $(\phi \equiv \psi)$.

Formulas are used as descriptions of objects of the universe. In particular, atomic formula of the form (a, v) is interpreted as a description of all objects having value v for attribute a.

The semantics for DL is given by a model. For DL, the *model* is KR-system $S = (U, A)$, which describes the meaning of symbols of predicates (a, v) in U, and if we properly interpret formulas in the model, then each formula becomes a meaningful sentence, expressing properties of some objects.

An object $x \in U$ *satisfies* a formula ϕ in $S = (U, A)$, denoted $x \models_S \phi$ or in shorts $x \models \phi$, iff the following conditions are satisfied:

1. $x \models (a, v)$ iff $a(x) = v$
2. $x \models \sim \phi$ iff $x \not\models \phi$
3. $x \models \phi \vee \psi$ iff $x \models \phi$ or $x \models \psi$
4. $x \models \phi \wedge \psi$ iff $x \models \phi$ and $x \models \psi$

The following are clear from the above truth definition:

5. $x \models \phi \rightarrow \psi$ iff $x \models \sim \phi \vee \psi$
6. $x \models \phi \equiv \psi$ iff $x \models \phi \rightarrow \psi$ and $x \models \psi \rightarrow \phi$

If ϕ is a formula, then the set $|\phi|_S$ defined as follows:

$|\phi|_s = \{x \in U \mid x \models_S \phi\}$

will be called the *meaning* of the formula ϕ in S.

Proposition 2.20 *The meaning of arbitrary formulas satisfies the following:*

$$|(a, v)|_S = \{x \in U \mid a(x) = v\}$$
$$|\sim \phi|_S = - |\phi|_S$$
$$|\phi \vee \psi|_S = |\phi|_S \cup |\psi|_S$$
$$|\phi \wedge \psi|_S = |\phi|_S \cap |\psi|_S$$
$$|\phi \rightarrow \psi|_S = - |\phi|_S \cup |\psi|_S$$
$$|\phi \equiv \psi|_S = (|\phi|_S \cap |\psi|_S) \cup (- |\phi|_S \cap - |\psi|_S)$$

Thus, the meaning of the formula ϕ is the set of all objects having the property expressed by the formula ϕ, or the meaning of the formula ϕ is the description in the KR-language of the set objects $|\phi|$.

A formula ϕ is said to be *true* in a KR-system S, denoted $\models_S \phi$, iff $|\phi|_S = U$, i.e., the formula is satisfied by all objects of the universe in the system S. Formulas ϕ and ψ are equivalent in S iff $|\phi|_S = |\psi|_S$.

Proposition 2.21 *The following are the simple properties of the meaning of a formula.*

$$\models_S \phi \textit{ iff } |\phi| = U$$
$$\models_S \sim\phi \textit{ iff } |\phi| = \emptyset$$
$$\phi \rightarrow \psi \textit{ iff } |\psi| \subseteq |\psi|$$
$$\phi \equiv \psi \textit{ iff } |\psi| = |\psi|$$

The meaning of the formula depends on the knowledge we have about the universe, i.e., on the knowledge representation system. In particular, a formula may be true in one knowledge representation system, but false in another one.

However, there are formulas which are true independent of the actual values of attributes appearing them. But, they depend only on their formal structure.

Note that in order to find the meaning of such a formula, one need not be acquainted with the knowledge contained in any specific knowledge representation system because their meaning is determined by its formal structure only.

Hence, if we ask whether a certain fact is true in light of our actual knowledge, it is sufficient to use this knowledge in an appropriate way. For formulas which are true (or not) in every possible knowledge representation system, we do not need in any particular knowledge, but only suitable logical tools.

To deal with deduction in DL, we need suitable axioms and inference rules. Here, axioms will correspond closely to axioms of classical propositional logic, but some specific axioms for the specific properties of knowledge representation systems are also needed. The only inference rule will be *modus ponens*.

We will use the following abbreviations:

$$\phi \wedge \sim\phi =_{\text{def}} 0$$
$$\phi \vee \sim\phi =_{\text{def}} 1$$

Obviously, $\models 1$ and $\models \sim 0$. Thus, 0 and 1 can be assumed to denote *falsity* and *truth*, respectively.

Formula of the form:

$$(a_1, v_1) \wedge (a_2, v_2) \wedge \ldots \wedge (a_n, v_n)$$

where $v_i \in V_a$, $P = \{a_1, a_2, \ldots, a_n\}$ and $P \subseteq A$ is called a *P-basic formula* or in short P-formula. Atomic formulas is called *A-basic formula* or in short basic formula.

Let $P \subseteq A$, ϕ be a P-formula and $x \in U$. If $x \models \phi$ then ϕ is called the *P-description* of x in S. The set of all A-basic formulas satisfiable in the knowledge representation system $S = (U, A)$ is called the *basic knowledge* in S.

We write $\sum_S(P)$, or in short $\sum(P)$, to denote the disjunction of all P-formulas satisfied in S. If $P = A$ then $\sum(A)$ is called the *characteristic formula* of S.

The knowledge representation system can be represented by a data table. And its columns are labelled by attributes and its rows are labelled by objects. Thus, each row in the table is represented by a certain A-basic formula, and the whole table is represented by the set of all such formulas. In DL, instead of tables, we can use sentences to represent knowledge.

There are specific axioms of DL:

1. $(a, v) \wedge (a, u) \equiv 0$ for any $a \in A, u, v \in V$ and $v \neq u$
2. $\bigvee_{v \in V_a} (a, v) \equiv 1$ for every $a \in A$
3. $\sim (a, v) \equiv \bigvee_{a \in V_a, u \neq v} (a, u)$ for every $a \in A$

The axiom (1) states that each object can have exactly one value of each attribute.

The axiom (2) assumes that each attribute must take one of the values of its domain for every object in the system.

The axiom (3) allows us to eliminate negation in such a way that instead of saying that an object does not possess a given property we can say that it has one of the remaining properties.

Proposition 2.22 *The following holds for DL:*

$$\models_S \sum_S(P) \equiv 1 \textit{ for any } P \subseteq A.$$

Proposition 2.22 means that the knowledge contained in the knowledge representation system is the whole knowledge available at the present stage. and corresponds to the so-called *closed world assumption* (CWA).

We say that a formula ϕ is *derivable* from a set of formulas Ω, denoted $\Omega \vdash \phi$, iff it is derivable from axioms and formulas of Ω by finite application of *modus ponens*. Formula ϕ is a theorem of DL, denoted $\vdash \phi$, if it is derivable from the axioms only. A set of formulas Ω is *consistent* iff the formula $\phi \wedge \sim \phi$ is not derivable from Ω.

Note that the set of theorems of DL is identical with the set of theorems of classical propositional logic with specific axioms (1)–(3), in which negation can be eliminated.

Formulas in the KR-language can be represented in a special form called *normal form*, which is similar to that in classical propositional logic.

Let $P \subseteq A$ be a subset of attributes and let ϕ be a formula in KR-language. We say that ϕ is in a *P-normal form* in S, in short in P-normal form, iff either ϕ is 0 or ϕ is 1, or ϕ is a disjunction of non-empty P-basic formulas in S. (The formula ϕ is non-empty if $|\phi| \neq \emptyset$).

A-normal form will be referred to as *normal form*. The following is an important property in the DL-language.

Table 2.1 *KR*-system 1

U	A	B	C
1	1	0	2
2	2	0	3
3	1	1	1
4	1	1	1
5	2	1	3
6	1	0	3

Proposition 2.23 *Let ϕ be a formula in DL-language and let P contain all attributes occurring in ϕ. Moreover, (1)–(3) and the formula $\sum_S(A)$. Then, there is a formula ψ in the P-normal form such that $\phi \equiv \psi$.*

Here is the example from Pawlak [26]. Consider the following *KR*-system.

The following $a_1b_0c_2, a_2b_0c_3, a_1b_1c_1, a_2b_1c_3, a_1b_0c_3$ are all basic formulas (basic knowledge) in the *KR*-system. For simplicity, we will omit the symbol of conjunction $\wedge$ in basic formulas.

The characteristic formula of the system is:

$$a_2b_0c_2 \vee a_2b_0c_3 \vee a_1b_1c_1 \vee a_2b_1c_3 \vee a_1b_0c_3$$

Here, we give the following meanings of some formulas in the system:

$$|a_1 \vee b_0c_2| = \{1, 2, 4, 6\}$$
$$|\sim (a_2b_1)| = \{1, 2, 3, 4, 6\}$$
$$|b_0 \to c_2| = \{1, 3, 4, 5\}$$
$$|a_2 \equiv b_0| = \{2, 3, 4\}$$

Below are given normal forms of formulas considered in the above example for *KR*-system 1:

$$a_1 \vee b_0c_2 = a_1b_0c_2 \vee a_1b_1c_1 \vee a_1b_0c_3$$
$$\sim (a_2b_1) = a_1b_0c_2 \vee a_2b_0c_3 \vee a_1b_1c_1 \vee a_1b_0c_3$$
$$b_0 \to c_2 = a_1b_0c_2 \vee a_1b_1c_1 \vee a_2b_1c_3$$
$$a_2 \equiv b_0 = a_2b_0c_1 \vee a_2b_0c_2 \vee a_2b_0c_3 \vee a_1b_1c_1 \vee a_1b_1c_2 \vee a_1b_1c_3$$

Examples of formulas in $\{a, b\}$-normal form are:

$$\sim (a_2b_1) = a_1b_0 \vee a_2b_0 \vee a_1b_1 \vee a_1b_0$$
$$a_1 \equiv b_0 = a_2b_0 \vee a_1b_1$$

The following is an example of a formula in $\{b, c\}$-normal form:

$$b_0 \to c_2 = b_0c_2 \vee b_1c_1 \vee b_1c_3$$

Thus, in order to compute the normal form of a formula, we have to transform by using propositional logic and the specific axioms for a given KR-system.

Any implication $\phi \to \psi$ is called a *decision rule* in the KR-language. ϕ and ψ are referred to as the *predecessor* and *successor* of $\phi \to \psi$, respectively.

If a decision rule $\phi \to \psi$ is true in S, we say that the decision rule is *consistent* in S; otherwise the decision rule is *inconsistent* in S.

If $\phi \to \psi$ is a decision rule and ϕ and ψ are P-basic and Q-basic formulas respectively, then the decision rule $\phi \to \psi$ is called a *PQ-basic decision rule* (in short PQ-rule).

A PQ-rule $\phi \to \psi$ is *admissible* in S if $\phi \wedge \psi$ is satisfiable in S.

Proposition 2.24 *A PQ-rule is true (consistent) in S iff all $\{P, Q\}$-basic formulas which occur in the $\{P, Q\}$-normal form of the predecessor of the rule, also occur in $\{P, Q\}$-normal form of the successor of the rule; otherwise the rule is false (inconsistent).*

The rule $b_0 \to c_2$ is false in the above table for KR-system 1, since the $\{b, c\}$-normal form of b_0 is $b_0c_2 \vee b_0c_3$, $\{b, c\}$-normal form of c_2 is b_0c_2, and the formula b_0c_3 does not occur in the successor of the rule.

On the other hand, the rule $a_2 \to c_3$ is true in the table, because the $\{a, c\}$-normal form of a_2 is a_2c_3, whereas the $\{a, c\}$-normal form of c_3 is $a_2c_3 \vee a_1c_3$.

Any finite set of decision rules in a DL-language is referred to as a *decision algorithm* in the DL-language. If all decision rules in a basic decision algorithm are PQ-decision rules, then the algorithm is said to be *PQ-decision algorithm*, or in short PQ-algorithm, and will be denoted by (P, Q).

A PQ-algorithm is *admissible* in S, if the algorithm is the set of all PQ-rules admissible in S.

A PQ-algorithm is *complete* in S, iff for every $x \in U$ there exists a PQ-decision rule $\phi \to \psi$ in the algorithm such that $x \models \phi \wedge \psi$ in S; otherwise the algorithm is *incomplete* in S.

A PQ-algorithm is *consistent* in S iff all its decision rules are consistent (true) in S; otherwise the algorithm is *inconsistent*.

Sometimes consistency (inconsistency) may be interpreted as *determinism* (*indeterminism*).

Given a KR-system, any two arbitrary, non-empty subset of attributes P, Q in the system determines uniquely a PQ-decision algorithm.

Consider the following KR-system from Pawlak [26].

Assume that $P = \{a, b, c\}$ and $Q = \{d, c\}$ are condition and decision attributes, respectively. Set P and Q uniquely associate the following PQ-decision algorithm with the table.

$a_1b_0c_2 \to d_1e_1$

$a_2b_1c_0 \to d_1e_0$

$a_2b_2c_2 \to d_0e_2$

$a_1b_2c_2 \to d_1e_1$

Table 2.2 *KR*-system 2

U	A	B	C	D	E
1	1	0	2	1	1
2	2	1	0	1	0
3	2	1	2	0	2
4	1	2	2	1	1
5	1	2	0	0	2

$a_1b_2c_0 \rightarrow d_0e_2$

If assume that $R = \{a, b\}$ and $T = \{c, d\}$ are *condition* and *decision attributes*, respectively, then the RT-algorithm determined by Table 2.2 is the following:

$a_1b_0 \rightarrow c_2d_1$
$a_2b_1 \rightarrow c_0d_1$
$a_2b_1 \rightarrow c_2d_0$
$a_1b_2 \rightarrow c_2d_1$
$a_1b_2 \rightarrow c_0d_0$

Of course, both algorithms are admissible and complete.

In order to check whether or not a decision algorithm is consistent, we have to check whether all its decision rules are true. The following proposition gives a much simpler method to solve this problem.

Proposition 2.25 *A PQ-decision rule $\phi \rightarrow \psi$ in a PQ-decision algorithm is consistent (true) in S iff for any PQ-decision rule $\phi' \rightarrow \psi'$ in PQ-decision algorithm, $\phi = \phi'$ implies $\psi = \psi'$.*

In Proposition 2.25, order of terms is important, since we require equality of expressions. Note also that in order to check whether or not a decision rule $\phi \rightarrow \psi$ is true we have to show that the predecessor of the rule (the formula ϕ) discerns the decision class ψ from the remaining decision classes of the decision algorithm in question. Thus, the concept of truth is somehow replaced by the concept of indiscernibility.

Consider the KR-system 2 again. With $P = \{a, b, c\}$ and Q as condition and decision attributes. Let us check whether the PQ-algorithm:

$a_1b_0c_2 \rightarrow d_1e_1$

$a_2b_1c_0 \rightarrow d_1e_0$

$a_2b_2c_2 \rightarrow d_0e_2$

$a_1b_2c_2 \rightarrow d_1e_1$

$a_1b_2c_0 \rightarrow d_0e_2$

is consistent or not. Because the predecessors of all decision rules in the algorithm are different (i.e., all decision rules are discernible by predecessors of all decision

Table 2.3 KR-system 2

U	A	B	C	D	E
1	1	0	2	1	1
4	1	2	2	1	1
2	2	1	0	1	0
3	2	1	2	0	2
5	1	2	0	0	2

rules in the algorithm), all decision rules in the algorithm are consistent (true) and consequently the algorithm is consistent.

This can also be seen directly from Table 2.3.

The RT-algorithm, where $R = \{a, b\}$ and $T\{c, d\}$

$$a_1b_0 \to c_2d_1$$
$$a_2b_1 \to c_0d_1$$
$$a_2b_1 \to c_2d_0$$
$$a_1b_2 \to c_2d_1$$
$$a_1b_2 \to c_0d_0$$

is inconsistent because the rules

$$a_2b_1 \to c_0d_1$$
$$a_2b_1 \to c_2d_0$$

have the same predecessors and different successors, i.e., we are unable to discern c_0d_1 and c_2d_0 by means of condition a_2b_1. Thus, both rules are inconsistent (false) in the KR-system. Similarly, the rules

$$a_1b_2 \to c_2d_1$$
$$a_1b_2 \to c_0d_0$$

are also inconsistent (false).

We turn to *dependency* of attributes. Formally, the dependency is defined as below. Let $K = (U, \mathbf{R})$ be a knowledge base and $\mathbf{P}, \mathbf{Q} \subseteq \mathbf{R}$.

(1) Knowledge $\mathbf{Q}$ *depends on knowledge* $\mathbf{P}$ iff $IND(\mathbf{P}) \subseteq IND(\mathbf{Q})$.
(2) Knowledge $\mathbf{P}$ and $\mathbf{Q}$ are *equivalent*, denoted $\mathbf{P} \equiv \mathbf{Q}$, if $\mathbf{P} \Rightarrow \mathbf{Q}$ and $\mathbf{Q} \Rightarrow \mathbf{P}$.
(3) Knowledge $\mathbf{P}$ and $\mathbf{Q}$ are *independent*, denoted $\mathbf{P} \not\equiv \mathbf{Q}$, iff neither $\mathbf{P} \Rightarrow \mathbf{Q}$ nor $\mathbf{Q} \Rightarrow \mathbf{P}$ hold.

Obviously, $\mathbf{P} \equiv \mathbf{Q}$ iff $IND(\mathbf{P}) \equiv IND(\mathbf{Q})$.

The dependency can be interpreted in different ways as Proposition 2.26 indicates:

Proposition 2.26 *The following conditions are equivalent:*

(1) $\mathbf{P} \Rightarrow \mathbf{Q}$
(2) $IND(\mathbf{P} \cup \mathbf{Q}) = INS(\mathbf{P})$
(3) $POS_{\mathbf{P}}(\mathbf{Q}) = U$
(4) $\underline{\mathbf{P}}X$ *for all* $X \in U/\mathbf{Q}$

where $\underline{\mathbf{P}}X$ *denotes* $\underline{IND(\mathbf{P})}/X$.

By Proposition 2.27, we can see the following: if **Q** depends on **P** then knowledge **Q** is superfluous within the knowledge base in the sense that the knowledge $\mathbf{P} \cup \mathbf{Q}$ and **P** provide the same characterization of objects.

Proposition 2.27 *If* **P** *is a reduct of* **Q**, *then* $\mathbf{P} \Rightarrow \mathbf{Q} - \mathbf{P}$ *and* $IND(\mathbf{P}) = IND(\mathbf{Q})$.

Proposition 2.28 *The following hold.*

(1) *If* **P** *is dependent, then there exists a subset* $\mathbf{Q} \subset \mathbf{P}$ *such that* **Q** *is a reduct of* **P**.
(2) *If* $\mathbf{P} \subseteq \mathbf{Q}$ *and* **P** *is independent, then all basic relations in* **P** *are pairwise independent.*
(3) *If* $\mathbf{P} \subseteq \mathbf{Q}$ *and* **P** *is independent, then every subset* **R** *of* **P** *is independent.*

Proposition 2.29 *The following hold:*

(1) *If* $\mathbf{P} \Rightarrow \mathbf{Q}$ *and* $\mathbf{P}' \supset \mathbf{P}$, *then* $\mathbf{P}' \Rightarrow \mathbf{Q}$.
(2) *If* $\mathbf{P} \Rightarrow \mathbf{Q}$ *and* $\mathbf{Q}' \subset \mathbf{Q}$, *then* $\mathbf{P} \Rightarrow \mathbf{Q}'$.
(3) $\mathbf{P} \Rightarrow \mathbf{Q}$ *and* $\mathbf{Q} \Rightarrow \mathbf{R}$ *imply* $\mathbf{P} \Rightarrow \mathbf{R}$.
(4) $\mathbf{P} \Rightarrow \mathbf{R}$ *and* $\mathbf{Q} \Rightarrow \mathbf{R}$ *imply* $\mathbf{P} \cup \mathbf{Q} \Rightarrow \mathbf{R}$.
(5) $\mathbf{P} \Rightarrow \mathbf{R} \cup \mathbf{Q}$ *imply* $\mathbf{P} \Rightarrow \mathbf{R}$ *and* $\mathbf{P} \cup \mathbf{Q} \Rightarrow \mathbf{R}$.
(6) $\mathbf{P} \Rightarrow \mathbf{Q}$ *and* $\mathbf{Q} \cup \mathbf{R} \Rightarrow \mathbf{T}$ *imply* $\mathbf{P} \cup R \Rightarrow \mathbf{T}$
(7) $\mathbf{P} \Rightarrow \mathbf{Q}$ *and* $\mathbf{R} \Rightarrow \mathbf{T}$ *imply* $\mathbf{P} \cup \mathbf{R} \Rightarrow \mathbf{Q} \cup \mathbf{T}$.

The derivation (dependency) can be partial, which means that only part of knowledge **Q** is derivable from knowledge **P**. We can define the partial derivability using the notion of the positive region of knowledge.

Let $K = (U, \mathbf{R})$ be the knowledge base and $\mathbf{P}, \mathbf{Q} \subset \mathbf{R}$. Knowledge **Q** depends in a degree k $(0 \leq k \leq 1)$ from knowledge **P**, in symbol $\mathbf{P} \Rightarrow_k \mathbf{Q}$, iff

$$k = \gamma_{\mathbf{P}}(\mathbf{Q}) = \frac{card(POS_{\mathbf{P}}(\mathbf{Q}))}{card(U)}$$

where $card$ denotes cardinality of the set.

If $k = 1$, we say that **Q** *totally depends from* **P**; if $0 < k < 1$, we say that **Q** *roughly (partially) depends from* **P**, and if $k = 1$ we say that **Q** is *totally independent from* **P**, If $\mathbf{P} \Rightarrow_1 \mathbf{Q}$, we shall also write $\mathbf{P} \Rightarrow \mathbf{Q}$.

The above ideas can also be interpreted as an ability to classify objects. More precisely, if $k = 1$, then all elements of the universe can be classified to elementary categories of $U/\mathbf{Q}$ by using knowledge $\mathbf{P}$.

Thus, the coefficient $\gamma_{\mathbf{P}}(\mathbf{Q})$ can be understood as a degree of dependency between, $\mathbf{Q}$ and $\mathbf{P}$. In other words, if we restrict the set of objects in the knowledge base to the set $POS_{\mathbf{P}}(\mathbf{Q})$, we would obtain the knowledge base in which $\mathbf{P} \Rightarrow \mathbf{Q}$ is a total dependency.

The measure k of dependency $\mathbf{P} \Rightarrow_k \mathbf{Q}$ does not capture how this partial dependency is actually distributed among classes of $U/\mathbf{Q}$. For example, some decision classes can be fully characterized by $\mathbf{P}$, whereas others may be characterised only partially.

We will also need a coefficient $\gamma(X) = card(\underline{\mathbf{P}}X)/card(X)$ where $X \in U/\mathbf{Q}$ which shows how many elements of each class of $U/\mathbf{Q}$ can be classified by employing knowledge $\mathbf{P}$.

Thus, the two numbers $\gamma(\mathbf{Q})$ and $\gamma(X)$, $x \in U/\mathbf{Q}$ give us full information about "classification power" of the knowledge $\mathbf{P}$ with respect to the classification $U/\mathbf{Q}$.

Proposition 2.30 *The following hold:*

(1) *If* $\mathbf{R} \Rightarrow_k \mathbf{P}$ *and* $\mathbf{Q} \Rightarrow_l \mathbf{P}$, *then* $\mathbf{R} \cup \mathbf{Q} \Rightarrow \mathbf{P}$, *for some* $m \geq max(k, l)$.

(2) *If* $\mathbf{R} \cup \mathbf{P} \Rightarrow_k \mathbf{Q}$ *then* $\mathbf{R} \Rightarrow_l \mathbf{Q}$ *and* $\mathbf{P} \Rightarrow_m \mathbf{Q}$, *for some* $l, m, \leq k$.

(3) *If* $\mathbf{R} \Rightarrow_k \mathbf{Q}$ *and* $\mathbf{R} \Rightarrow_l \mathbf{P}$, *then* $\mathbf{R} \Rightarrow_m \mathbf{Q} \cup \mathbf{P}$, *for some* $m \leq max(k, l)$.

(4) *If* $\mathbf{R} \Rightarrow_k \mathbf{Q} \cup \mathbf{P}$, *then* $\mathbf{R} \Rightarrow_l \mathbf{Q}$ *and* $\mathbf{R} \Rightarrow_m \mathbf{P}$, *for some* $l, m \geq k$.

(5) *If* $\mathbf{R} \Rightarrow_k \mathbf{P}$ *and* $\mathbf{P} \Rightarrow_l \mathbf{Q}$, *then* $\mathbf{R} \Rightarrow_m \mathbf{Q}$, *for some* $m \geq l + k - 1$.

Here, we return to the decision algorithm for dependency. We say that the set of attributes Q depends *totally*, (or in short depends) on the set of attributes P in S, if there exists a consistent PQ-algorithm in S. If Q depends on P in S, we write $P \Rightarrow_S S$, or in short $P \Rightarrow Q$.

We can also define partial dependency of attributes. We say that the set of attributes Q *depends partially* on the set of attributes P in S, if there exists an inconsistent PQ-algorithm in S.

The degree of *dependency* between attributes can be defined. Let (P, Q) be a PQ-algorithm in S. By a *positive region* of the algorithm (P, Q), denoted $POS(P, Q)$, we mean the set of all consistent (true) PQ-rules in the algorithm.

The positive region of the decision algorithm (P, Q) is the consistent part (possibly empty) of the inconsistent algorithm. Obviously, a PQ-algorithm is inconsistent iff $POS(P, Q) \neq (P, Q)$ or what is the same $card(POS(P, Q)) \neq card(P, Q)$.

With every PQ-decision algorithm, we can associate a number $k = card(POS(P, Q))/card(P, Q)$, called the *degree of consistency*, of the algorithm, or in short the degree of the algorithm, and we say that the PQ-algorithm has the degree (of consistency) k.

Obviously, $0 \leq k \leq 1$. If a PQ-algorithm has degree k, we can say that the set of attributes Q *depend in degree* k on the set of attributes P, denoted $P \Rightarrow_k Q$.

Naturally, the algorithm is consistent iff $k = 1$; otherwise, i.e., if $k \neq 1$, the algorithm. All these concepts are the same as in those discussed above. Note that in the consistent algorithm all decisions are uniquely determined by conditions in the decision algorithm.

In other words, this means that all decisions in a consistent algorithm are discernible by means of conditions available in the decision algorithm.

Decision logic provides a simple means for reasoning about knowledge only by using propositional logic, and is suitable to some applications. Note here that the so-called *decision table* can serve as a KR-system.

However, the usability of decision logic seems to be restrictive. In other words, it is far from a general system for reasoning in general. In this book, we will lay general frameworks for reasoning based on rough set theory.

2.2 Decision Tables

Decision tables can be seen as a special, important class of knowledge representation systems, and can be used for applications. Let $K = (U, A)$ be a knowledge representation system (*KR*-system) and $C, D \subset A$ be two subsets of attributes, called *condition* and *decision* attributes, respectively.

2.2.1 Definitions

KR-system with distinguished condition and decision attributes is called a *decision table*, denoted $T = (U, A, C, D)$ or in short DC. Equivalence classes of the relations $IND(C)$ and $IND(D)$ are called *condition* and *decision classes*, respectively.

With every $x \in U$, we associate a function $d_x : A \rightarrow V$, such that $d_x(a) = a(x)$ for every $a \in C \cup D$; the function d_x is called a *decision rule* (in T), and x is referred as a *label* of the decision rule d_x.

Note that elements of the set U in a decision table do not represent in general any real objects, but are simple identifiers of decision rules.

If d_x is a decision rule, then the restriction of d_x to C, denoted $d_x \mid C$, and the restriction of d_x to D, denoted $d_x \mid D$ are called *conditions* and *decisions* (actions) of d_x, respectively.

The decision rule d_x is *consistent* (in T) if for every $y \neq x, d_x \mid C = d_y \mid C$ implies $d_x \mid D = d_y \mid D$; otherwise the decision rule is *inconsistent*.

A decision table is *consistent* if all its decision rules are consistent; otherwise the decision table is *inconsistent*. Consistency (inconsistency) sometimes may be interpreted as determinism (non-determinism).

Proposition 2.31 *A decision table $T = (U, A, C, D)$ is consistent iff $C \Rightarrow D$.*

Table 2.4 Decision Table 2.1

U	A	B	C	D	E
1	1	0	2	2	0
2	0	1	1	1	2
3	2	0	0	1	1
4	1	1	0	2	2
5	1	0	2	0	1
6	2	2	0	1	1
7	2	1	1	1	2
8	0	1	1	0	1

From Proposition 2.31, it follows that the practical method of checking consistency of a decision table is by simply computing the degree of dependency between condition and decision attributes. If the degree of dependency equals to 1, then we conclude that the table is consistent; otherwise it is inconsistent.

Proposition 2.32 *Each decision table $T = (U, A, C, D)$ can be uniquely decomposed into two decision tables $T_1 = (U, A, C, D)$ and $T_2 = (U, A, C, D)$ such that $C \Rightarrow_1 D$ in T_1 and $C \Rightarrow_0 D$ in T_2 such that $U_1 = POS_C(D)$ and $U_2 = \bigcup_{X \in U/IND(D)} BN_C(X)$.*

Proposition 2.32 states that we can decompose the table into two subtables; one totally inconsistent with dependency coefficient equal to 0, and the second entirely consistent with the dependency equal to 1. This decomposition however is possible only if the degree of dependency is greater than 0 and different from 1.

Consider Table 2.4 from Pawlak [26].

Assume that a, b and c are condition attributes, and d and e are decision attributes. In this table, for instance, the decision rule 1 is inconsistent, whereas the decision rule 3 is consistent. By Proposition 2.32, we can decompose Decision Table 2.1 into the following two tables:

Table 2.2 is consistent, whereas Table 2.3 is totally inconsistent, which means all decision rules in Table 2.2 are consistent, and in Table 2.3 all decision rules are inconsistent.

Simplification of decision tables is very important in many applications, e.g. software engineering. An example of simplification is the reduction of condition attributes in a decision table.

In the reduced decision table, the same decisions can be based on a smaller number of conditions. This kind of simplification eliminates the need for checking unnecessary conditions.

2.2.2 Simplification

Pawlak proposed *simplification* of decision tables which includes the following steps:

(1) Computation of reducts of condition attributes which is equivalent to elimination of some column from the decision table.
(2) Elimination of duplicate rows.
(3) Elimination of superfluous values of attributes.

Thus, the method above consists in removing superfluous condition attributes (columns), duplicate rows and, in addition to that, irrelevant values of condition attributes.

By the above procedure, we obtain an "incomplete" decision table, containing only those values of condition attributes which are necessary to make decisions. According to our definition of a decision table, the incomplete table is not a decision table and can be treated as an abbreviation of such a table.

For the sake of simplicity, we assume that the set of condition attributes is already reduced, i.e., there are not superfluous condition attributes in the decision table.

With every subset of attributes $B \subseteq A$, we can associate partition $U/IND(B)$ and consequently the set of conditions and decision attributes define partitions of objects into condition and decision classes.

We know that with every subset of attributes $B \subseteq A$ and object x we may associate set $[x]_B$, which denotes an equivalence class of the relation $IND(B)$ containing an object x, i.e., $[x]_B$ is an abbreviation of $[x]_{IND(B)}$.

Thus, with any set of condition attributes C in a decision rule d_x we can associate set $[x]_C = \bigcap_{a \in C}[x]_a$. But, each set $[x]_a$ is uniquely determined by attribute value $a(x)$. Hence, in order to remove superfluous values of condition attributes, we have to eliminate all superfluous equivalence classes $[x]_a$ from the equivalence class $[x]_C$. Thus, problems of elimination of superfluous values of attributes and elimination of corresponding equivalence classes are equivalent.

Consider the following decision table from Pawlak [26].

Here, a, b and c are condition attributes and e is a decision attribute.

It is easy to compute that the only e-dispensable condition attribute is c; consequently, we can remove column c in Table 2.4, which yields Table 2.5:

In the next step, we have to reduce superfluous values of condition attributes in every decision rule. First, we have to compute core values of condition attributes in every decision rule.

Table 2.5 Decision Table 2.2

U	A	B	C	D	E
3	2	0	0	1	1
4	1	1	0	2	2
6	2	2	0	1	1
7	2	1	1	1	2

Table 2.6 Decision Table 2.3

U	A	B	C	D	E
1	1	0	2	2	0
2	0	1	1	1	2
5	1	0	2	0	1
8	0	1	1	0	1

Here, we compute the core values of condition attributes for the first decision rule, i.e., the core of the family of sets

$$\mathbf{F} = \{[1]_a, [1]_b, [1]_d\} = \{\{1, 2, 4, 5\}, \{1, 2, 3\}, \{1, 4\}\}$$

From this we have:

$$[1]_{\{a,b,d\}} = [1]_a \cap [1]_b \cap [1]_d = \{1, 2, 4, 5\} \cap \{1, 2, 3\} \cap \{1, 4\} = \{1\}.$$

Moreover, $a(1) = 1$, $b(1) = 0$ and $d(1) = 1$. In order to find dispensable categories, we have to drop one category at a time and check whether the intersection of remaining categories is still included in the decision category $[1]_e = \{1, 2\}$, i.e.,

$$[1]_b \cap [1]_d = \{1, 2, 3\} \cap \{1, 4\} = \{1\}$$
$$[1]_a \cap [1]_d = \{1, 2, 4, 5\} \cap \{1, 4\} = \{1, 4\}$$
$$[1]_a \cap [1]_b = \{1, 2, 4, 5\} \cap \{1, 2, 3\} = \{1, 2\}$$

This means that the core value is $b(1) = 0$. Similarly, we can compute remaining core values of condition attributes in every decision rule and the final results are represented in Table 2.6.

Then, we can proceed to compute value reducts. As an example, let us compute value reducts for the first decision rule of the decision table.

According to the definition of it, in order to compute reducts of the family $\mathbf{F} = \{[1]_a, [1]_b, [1]_d\} = \{\{1, 2, 3, 5\}, \{1, 2, 3\}, \{1, 4\}\}$, we have to find all subfamilies $\mathbf{G} \subseteq \mathbf{F}$ such that $\bigcap \mathbf{G} \subseteq [1]_e = \{1, 2\}$. There are four following subfamilies of $\mathbf{F}$:

$$[1]_b \cap [1]_d = \{1, 2, 3\} \cap \{1, 4\} = \{1\}$$
$$[1]_a \cap [1]_d = \{1, 2, 4, 5\} \cap \{1, 4\} = \{1, 4\}$$
$$[1]_a \cap [1]_b = \{1, 2, 4, 5\} \cap \{1, 2, 3\} = \{1\}$$

and only two of them

$$[1]_b \cap [1]_d = \{1, 2, 3\} \cap \{1, 4\} = \{1\} \subseteq [1]_e = \{1, 2\}$$
$$[1]_a \cap [1]_b = \{1, 2, 4, 5\} \cap \{1, 2, 3\} = \{1\} \subseteq [1]_e = \{1, 2\}$$

are reducts of the family $\mathbf{F}$. Hence, we have two values reducts: $b(1) = 0$ and $d(1) = 1$ or $a(1) = 1$ and $b(1) = 0$. This means that the attribute values of attributes a and b or d and e are characteristic for decision class 1 and do not occur in any other

Table 2.7 Decision Table 2.4

U	A	B	C	D	E
1	1	0	0	1	1
2	1	0	0	0	1
3	0	0	0	0	0
4	1	1	0	1	0
5	1	1	0	2	2
6	2	1	0	2	2
7	2	2	2	2	2

Table 2.8 Decision Table 2.5

U	A	B	D	E
1	1	0	1	1
2	1	0	0	1
3	0	0	0	0
4	1	1	1	0
5	1	1	2	2
6	2	1	2	2
7	2	2	2	2

Table 2.9 Decision Table 2.6

U	A	B	D	E
1	–	0	–	1
2	1	–	–	1
3	0	–	–	0
4	–	1	1	0
5	–	–	2	2
6	–	–	–	2
7	–	–	–	2

decision classes in the decision table. We see also that the value of attribute b is the intersection of both value reducts, $b(1) = 0$, i.e., it is the core value.

In Table 2.7, we list value reducts for all decision rules in Table 2.1.

Seen from Decision Table 2.7, for decision rules 1 and 2 we have two value reducts of condition attributes. Decision rules 3, 4 and 5 have only one value reducts of condition attributes for each decision rule row. The remaining decision rules 6 and 7 contain two and three value reducts, respectively.

Hence, there are two reduced form of decision rule 1 and 2, decision rule 3, 4 and 5 have only one reduced form each, decision rule 6 has two reducts and decision rule 7 has three reducts.

Thus, there are $4 \times 2 \times 3 = 24$ (not necessarily different) solutions to our problem. One such solution is presented in Decision Table 2.8.

Another solution is shown in Decision Table 2.9.

Table 2.10 Decision Table 2.7

U	A	B	D	E
1	1	0	×	1
1′	×	0	1	1
2	1	0	×	1
2′	1	×	0	1
3	0	×	×	0
4	×	1	1	0
5	×	×	2	2
6	×	×	2	2
6′	2	×	×	2
7	×	×	2	2
7′	×	2	×	2
7″	2	×	×	2

Table 2.11 Decision Table 2.8

U	A	B	D	E
1	1	0	×	1
2	1	×	0	1
3	0	×	×	0
4	×	1	1	0
5	×	×	2	2
6	×	×	2	2
7	2	×	×	2

Because decision rules 1 and 2 are identical, and so are rules 5, 6 and 7, we can represent Decision Table 2.10:

In fact, enumeration of decision rules is not essential, so we can enumerate them arbitrary and we get as a final result Decision Table 2.11:

This solution is referred to as *minimal*. The presented method of decision table simplification can be named *semantic*, since it refers to the meaning of the information contained in the table. Another decision table simplification is also possible and can be named *syntactic*. It is described within the framework of decision logic (Table 2.12).

To simplify a decision table, we should first find reducts of condition attributes, remove duplicate rows and then find value-reducts of condition attributes and again, if necessary, remove duplicate rows (Table 2.13).

This method leads to a simple algorithm for decision table simplification, Note that a subset of attributes may have more than one reduct (relative reduct). Thus, the simplification od decision table does not yield unique results. Some decision tables possibly can be optimized according to preassumed criteria (Table 2.14).

Table 2.12 Decision Table 2.9

U	A	B	D	E
1	1	0	×	1
2	1	0	×	1
3	0	×	×	0
4	×	1	1	0
5	×	×	2	2
6	×	×	2	2
7	×	×	2	2

Table 2.13 Decision Table 2.10

U	A	B	D	E
1, 2	1	0	×	1
3	0	×	×	0
4	×	1	1	0
5, 6, 7	×	×	2	2

Table 2.14 Decision Table 2.11

U	A	B	D	E
1	1	0	×	1
2	0	×	×	0
3	×	1	1	0
4	×	×	2	2

We have finished the presentation of some topics in rough set theory. Pawlak also established other formal results about rough sets and discussed advantages of rough set theory. We here omit these issues; see Pawlak [26].

2.3 Non-Classical Logics

Non-classical logic is a logic which differs from classical logic in some points especially aims to extend the interpretation of truth based on bivalence. There are many systems of non-classical logic in the literature. Furthermore, some non-classical logics are closely tied with foundations of rough set theory.

There are two types of non-classical logics. The first type is considered as an extension of classical logic. It extends classical logic with new features. For instance, modal logic adds modal operators to classical logic.

The second type is an alternative to classical logic called *many-valued logic* and it therefore lacks some of the features of classical logic.

For example, many-valued logic is based on many truth-values, whereas classical logic uses two truth-values, i.e. true and false. These two types of non-classical logics are conceptually different and their uses heavily depend on applications.

In some cases, they can provide more promising results than classical logic. In the following, we provide the basics of modal, many-valued logic., intuitionistic, and paraconsistent logic.

In this section, we briefly describe some non-classical logics. Some of them are clearly related to granular reasoning discussed in later chapters. For more on non-classical logics, see Akama [1] and Priest [27, 28].

2.3.1 Modal Logic

Modal logic extends classical logic with modal operators to represent intensional concepts. which cannot be properly treated by classical logic. So a new mechanism for intensionality should be extended. This can be described by a modal operator.

Generally, $\Box$ (necessity) and $\Diamond$ (possibility) are used as modal operators. A formula of the form $\Box A$ reads "A is necessarily true" and $\Diamond A$ "A is possibly true", respectively. These are dual in the sense that $\Box A \leftrightarrow \neg \Diamond \neg A$.

In addition, reading modal operators differently, we can obtain other intensional logics capable of formalizing some intensional concepts. Currently, many variants of modal logics are known, e.g. *tense logic*, *epistemic logic*, *doxastic logic*, *deontic logic*, *dynamic logic*, *intensional logic*, etc.

For example, Prior [29] developed tense logic for tenses in natural language. Hintikka [13] formalized logics for knowledge and beliefs, i.e., epistemic and doxastic logics.

Here, we present proof and model theory for modal logic. We only treat the most basic modal logics and Kripke model as its semantics for the later extension for granular computing. The language of the minimal modal logic denoted **K** is the classical propositional logic **CPC** with the necessity operator $\Box$. A Hilbert system for **K** is formalized as follows:

Modal Logic K

Axiom

(CPC) Axiom of **CPC**

(K) $\Box(A \to B) \to (\Box A \to \Box B)$

Rules of Inference

(MP) $\vdash A, \vdash A \to B \Rightarrow \vdash B$

(NEC) $\vdash A \Rightarrow \vdash \Box A$

Here, $\vdash A$ means that A is provable in **K**. (NEC) is called the *necessitation*. The notion of proof is defined as usual.

Systems of *normal modal logic* can be obtained by adding extra axioms which describe properties of modality. Some of the important axioms are listed as follows:

(D) $\Box A \rightarrow \Diamond A$
(T) $\Box A \rightarrow A$
(B) $A \rightarrow \Box\Diamond A$
(4) $\Box A \rightarrow \Box\Box A$
(5) $\Diamond A \rightarrow \Box\Diamond A$

The name of normal modal logic is systematically given by the combination of axioms. For instance, the extension of **K** with the axiom (D) is called KD. However, such systems traditionally have another names as follows:

D = KD
T = KT
B = KB
S4 = KT4
S5 = KT5.

Before the 1960s, the study of modal logic was mainly proof-theoretical due to the lack of model theory. A semantics of modal logic has been studied by Kripke and it is now called Kripke semantics [18–20].

Kripke semantics uses a possible world to interpret modal operators. Intuitively, the interpretation of $\Box A$ says that A is true in all possible worlds. Possible worlds are linked with the actual world by means of the accessibility relation.

A *Kripke model* for the normal modal logic **K** is defined as a triple $M = \langle W, R, V \rangle$, where W is a non-empty set of possible worlds, R is an *accessibility relation* on $W \times W$, and V is a valuation function: $W \times PV \rightarrow \{0, 1\}$. We here denote by PV a set of propositional variables. $F = \langle W, R \rangle$ is called a *frame* or a Kripke frame.

We write $M, w \models A$ to mean that a formula A is true at a world w in the model M. Let p be a propositional variable and false be absurdity. Then, the semantic relation $\models$ can be defined as follows:

$$M, w \models p \Leftrightarrow V(w, p) = 1$$
$$M, w \not\models \textit{false}$$
$$M, w \models \neg A \Leftrightarrow M, w \not\models A$$
$$M, w \models A \wedge B \Leftrightarrow M, w \models A \text{ and } M, w \models B$$
$$M, w \models A \vee B \Leftrightarrow M, w \models A \text{ or } M, w \models B$$
$$M, w \models A \rightarrow B \Leftrightarrow M, w \models A \Leftrightarrow M, w \models B$$
$$M, w \models \Box A \Leftrightarrow \forall v (wRv \Leftrightarrow M, v \models A)$$
$$M, w \models \Diamond A \Leftrightarrow \exists v (wRv \text{ and } M, v \models A)$$

Here, there are no restrictions on the property of R. We say that a formula A is valid in the modal logic S, written $M \models_S A$, just in case $M, w \models A$ for every world

Table 2.15 Kripke models

Axiom	Condition on R
(K)	No conditions
(D)	$\forall w \exists v(wRv)$ (serial)
(T)	$\forall w(wRw)$ (reflexive)
(4)	$\forall wvu(wRv$ and $vRu \Leftrightarrow wRu)$ (transitive)
(5)	$\forall wvu(wRv$ and $wRu \Leftrightarrow vRu)$ (euclidean)

w and every model M. We know that the minimal modal logic **K** is complete (Table 2.15).

Theorem 2.1 $\vdash_{\mathbf{K}} A \Leftrightarrow \models_{\mathbf{K}} A$.

By imposing some restrictions on the accessibility relation R, we can give Kripke models for various normal modal logics. The correspondences of axioms and conditions on R are given as follows:

For example, the accessibility relation in a Kripke model for modal logic **S4** is reflexive and transitive since it needs axioms (K), (T) and (4). The completeness results of several modal logics have been established in Hughes and Cresswell [14, 15] for details.

If we read modal operators differently, then other types of modal logics listed above can be obtained. These logics can deal with various problems, and modal logic is of special importance to applications.

2.3.2 *Many-Valued Logic*

Many-valued logic, also known as *multiple-valued logic*, is a family of logics that have more than two truth-values. Namely, many-valued logics can express other possibilities in addition to truth and falsity. The idea of many-valued logic is implicit in Aristotle's thinking concerning *future contingents*.

Now, many-valued logics are widely used for various applications, including hardware circuits. It is also noted that the so-called fuzzy logic is classified as a many-valued (infinite-valued) logic.

We start with the exposition of *three-valued logic*. The first serious attempt to formalize a three-valued logic has been done by Łukasiewicz in [22]. His system is now known as Łukasiewicz's three-valued logic, denoted $\mathbf{L}_3$, in which the third truth-valued reads "indeterminate" or "possible".

Łukasiewicz considered that future contingent propositions should receive the third truth-value denoted by I, which is neither true nor false, although his interpretation is controversial.

Table 2.16 Truth-value tables of $\mathbf{L_3}$

A	$\sim A$
T	F
I	I
F	T

A	B	$A \wedge B$	$A \vee B$	$A \rightarrow_L B$
T	T	T	T	T
T	F	F	T	F
T	I	I	T	I
F	T	F	F	T
F	F	F	F	T
F	I	F	I	T
I	T	I	T	T
I	F	F	I	I
I	I	I	I	T

The language of $\mathbf{L_3}$ comprises conjunction ($\wedge$), disjunction ($\vee$), implication ($\rightarrow_L$) and negation ($\sim$). The semantics for many-valued logics can be usually given by using the truth-value tables. The truth-value tables of $\mathbf{L_3}$ are as follows (Table 2.16):

Here, we should note that both the law of excluded middle $A \vee \sim A$ and the law of non-contradiction $\sim (A \wedge \sim A)$, which are important principles of classical logic, do not hold. In fact, these receive I when the truth-values of compound formulas are I.

A Hilbert system for $\mathbf{L_3}$ is as follows:

Lukasiewicz's Three-Valued Logic $\mathbf{L_3}$

Axiom

(L1) $A \rightarrow (B \rightarrow A)$

(L2) $(A \rightarrow B) \rightarrow ((B \rightarrow C) \rightarrow (A \rightarrow C))$

(L3) $((A \rightarrow \sim A) \rightarrow A) \rightarrow A$

(L4) $(\sim A \rightarrow \sim B) \rightarrow (B \rightarrow A)$

Rules of Inference

(MP) $\vdash A, \vdash A \rightarrow B \Rightarrow \vdash B$

Here, $\wedge$ and $\vee$ are defined by means of $\sim$ and $\rightarrow_L$ in the following way.

$$A \vee B =_{\text{def}} (A \rightarrow B) \rightarrow B$$

$$A \wedge B =_{\text{def}} \sim (\sim A \vee \sim B)$$

Table 2.17 Truth-value tables of $\mathbf{K_3}$

A	$\sim A$
T	F
I	I
F	T

A	B	$A \wedge B$	$A \vee B$	$A \rightarrow_K B$
T	T	T	T	T
T	F	F	T	F
T	I	I	T	I
F	T	F	F	T
F	F	F	F	T
F	I	F	I	T
I	T	I	T	T
I	F	F	I	I
I	I	I	I	I

Kleene also proposed three-valued logic $\mathbf{K_3}$ in connection with *recursive function theory*; see Kleene [17]. $\mathbf{K_3}$ differs from $\mathbf{L_3}$ in its interpretation of implication $\rightarrow_K$. The truth-value tables of $\mathbf{K_3}$ are given as Table 2.17.

In $\mathbf{K_3}$, the third truth-value reads "undefined". Consequently, $\mathbf{K_3}$ can be applied to theory of programs. There are no tautologies in $\mathbf{K_3}$, thus implying that we cannot provide a Hilbert system for $\mathbf{K_3}$. For example, it was used for the semantics for logic programs with negation as failure (NAF) by Fitting [8] and Kunen [21].

$\mathbf{K_3}$ is usually called *Kleene's strong three-valued logic*. In the literature, *Kleene's weak three-valued logic* also appears, in which a formula evaluates as I if any compound formula evaluates as I. Kleene's weak three-valued logic is equivalent to Bochvar's three-valued logic.

Four-valued logic is suited as a logic for a computer which must deal with incomplete and inconsistent information. Belnap introduced a four-valued logic which can formalize the internal states of a computer; see Belnap [6, 7].

There are four states, i.e. (T), (F), (None) and (Both), to recognize an input in a computer. Based on these states, a computer can compute suitable outputs.

(*T*) a proposition is true.

(*F*) a proposition is false.

(*N*) a proposition is neither true nor false.

(*B*) a proposition is both true and false.

Fig. 2.1 Approximation lattice

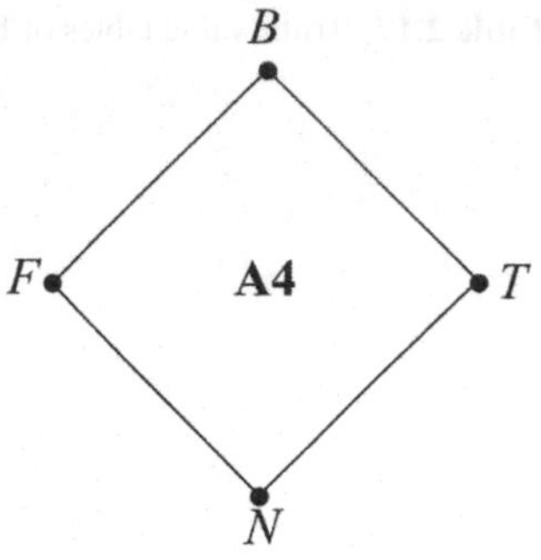

Fig. 2.2 Logical lattice

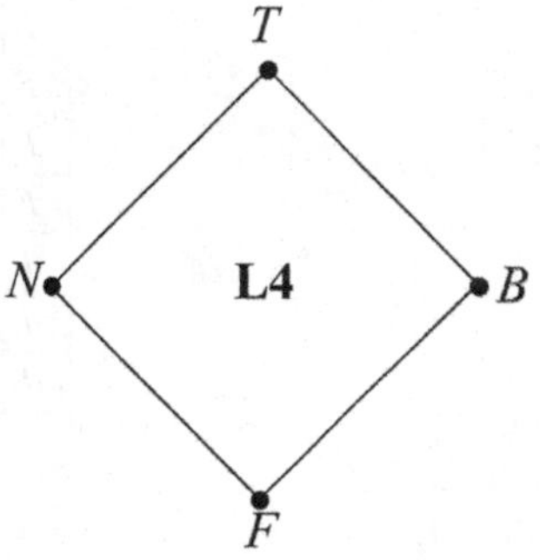

Here, (N) and (B) abbreviate (*None*) and (*Both*), respectively. From the above, (N) corresponds to incompleteness and (B) inconsistency. Four-valued logic can be thus seen as a natural extension of three-valued logic.

In fact, Belnap's four-valued logic can model both incomplete information (N) and inconsistent information (B). Belnap proposed two four-valued logics **A4** and **L4**.

The former can cope only with atomic formulas, whereas the latter can handle compound formulas. **A4** is based on the approximation lattice depicted as Fig 2.1.

Here, B is the least upper bound and N is the greatest lower bound with respect to the ordering $\leq$. **L4** is based on the logical lattice depicted as Fig 2.2. **L4** has logical symbols; $\sim$, $\wedge$, $\vee$, and is based on a set of truth-values $\mathbf{4} = \{T, F, N, B\}$. One of the features of **L4** is the monotonicity of logical symbols (Table 2.18).

Let f be a logical operation. It is said that f is monotonic iff $a \subseteq b \Rightarrow f(a) \subseteq f(b)$. To guarantee the monotonicity of conjunction and disjunction, they must satisfy the following:

$$a \wedge b = a \Leftrightarrow a \vee b = b,$$
$$a \wedge b = b \Leftrightarrow a \vee b = a.$$

The truth-value tables of **L4** are as follows:

Table 2.18 Truth-value tables of **L4**

	N	F	T	B
$\sim$	B	T	F	N

$\wedge$	N	F	T	B
N	N	F	N	F
F	F	F	F	F
T	N	F	T	B
B	F	F	B	B

$\vee$	N	F	T	B
N	N	N	T	T
F	N	F	T	B
T	T	T	T	T
B	T	B	T	B

Belnap gave a semantics for the language with the above logical symbols. A setup is a mapping a set of atomic formulas Atom to the set **4**. Then, the meaning of formulas of **L4** is defined as follows:

$$s(A \wedge B) = s(A) \wedge s(B)$$
$$s(A \vee B) = s(A) \vee s(B)$$
$$s(\sim A) = \sim s(A)$$

Further, Belnap defined an entailment relation $\rightarrow$ as follows:

$$A \rightarrow B \Leftrightarrow s(A) \leq s(B)$$

for all set-ups s.

The entailment relation $\rightarrow$ can be axiomatized as follows:

$$(A_1 \wedge \ldots \wedge A_m) \rightarrow (B_1 \vee \ldots \vee B_n) \; (A_i \text{ shares some } B_j)$$
$$(A \vee B) \rightarrow C \leftrightarrow (A \rightarrow C) \text{ and } (B \rightarrow C)$$
$$A \rightarrow B \Leftrightarrow \sim B \rightarrow \sim A$$
$$A \vee B \leftrightarrow B \vee A, \; A \wedge B \leftrightarrow B \wedge A$$
$$A \vee (B \vee C) \leftrightarrow (A \vee B) \vee C$$
$$A \wedge (B \wedge C) \leftrightarrow (A \wedge B) \wedge C$$
$$A \wedge (B \vee C) \leftrightarrow (A \wedge B) \vee (A \wedge C)$$
$$A \vee (B \wedge C) \leftrightarrow (A \vee B) \wedge (A \vee C)$$
$$(B \vee C) \wedge A \leftrightarrow (B \wedge A) \vee (C \wedge A)$$
$$(B \wedge C) \vee A \leftrightarrow (B \vee A) \wedge (C \vee A)$$
$$\sim\sim A \leftrightarrow A$$
$$\sim (A \wedge B) \leftrightarrow \sim A \vee \sim B, \; \sim (A \vee B) \leftrightarrow \sim A \wedge \sim B$$
$$A \rightarrow B, B \rightarrow C \Leftrightarrow A \rightarrow C$$

$A \leftrightarrow B, B \leftrightarrow C \Leftrightarrow A \leftrightarrow C$

$A \rightarrow B \Leftrightarrow A \leftrightarrow (A \wedge B) \Leftrightarrow (A \vee B) \leftrightarrow B$

Note here that $(A \wedge \sim A) \rightarrow B$ and $A \rightarrow (B \vee \sim B)$ cannot be derived in this axiomatization. It can be shown that the logic given above is closely related to the so-called relevant logic of Anderson and Belnap [2]. In fact, Belnap's four-valued logic is equivalent to the system of tautological entailment.

Infinite-valued logic is a many-valued logic having infinite truth-values in [0, 1]. *Fuzzy logic* and *probabilistic logic* belong to this family.

Łukasiewicz introduced infinite-valued logic $\mathbf{L}_\infty$ in 1930; see Łukasiewicz [23]. Its truth-value tables can be generated by the following matrix:

$$
\begin{aligned}
&|\sim A| = 1 - |A| \\
&|A \vee B| = max(|A|, |B|) \\
&|A \wedge B| = min(|A|, |B|) \\
&|A \rightarrow B| = 1 \qquad (|A| \leq |B|) \\
&\qquad\quad = 1 - |A| + |B| \ (|A| > |B|)
\end{aligned}
$$

A Hilbert system for L_∞ is as follows:

Łukasiewicz's Infinite-Valued logic L_∞
Axioms
(IL1) $A \rightarrow (B \rightarrow A)$
(IL2) $(A \rightarrow B) \rightarrow ((B \rightarrow C) \rightarrow (A \rightarrow C))$
(IL3) $((A \rightarrow B) \rightarrow B) \rightarrow ((B \rightarrow A) \rightarrow A)$
(IL4) $(\sim A \rightarrow \sim B) \rightarrow (B \rightarrow A)$
(IL5) $((A \rightarrow B) \rightarrow (B \rightarrow A)) \rightarrow (B \rightarrow A)$
Rules of Inference
(MP) $\vdash A, \vdash A \rightarrow B \Rightarrow \vdash B$

Since (IL5) is derived from other axioms, it can be deleted. It is known that L_∞ was used as the basis of *fuzzy logic* based on *fuzzy set* due to Zadeh [33].

Fuzzy logic is a logic of vagueness and can find many applications to various areas, in particular, engineering. Since the 1990s, a lot of important work has been done for foundations for fuzzy logic (cf. Hajek [12]). However, it is noted that fuzzy logic refers to logics which can deal with vagueness in a very wide sense.

2.3.3 Bilattice Logics

Fitting [9, 10] studied bilattice, which is the lattice **4** with two kinds of orderings, in connection with the semantics of logic programs. Bilattices introduce non-standard logical connectives.

Fig. 2.3 The bilattice *FOUR*

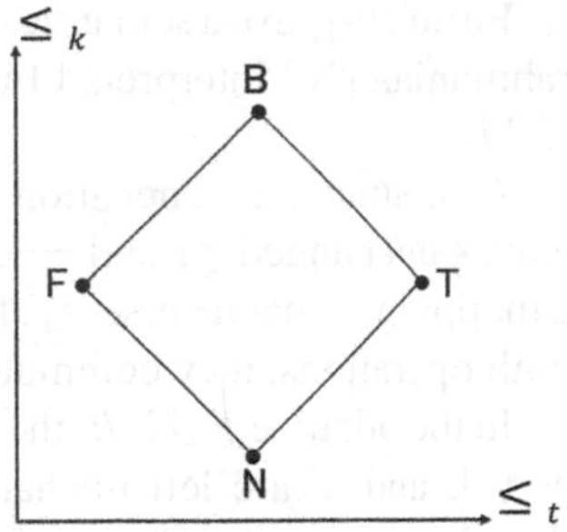

A *bilattice* was originally introduced by Ginsberg [11] for the foundations of reasoning in AI, which has two kinds of orderings, i.e., truth ordering and knowledge ordering.

Later, it was extensively studied by Fitting in the context of logic programming in [8] and of the theory of truth in [9]. In fact, bilattice-based logics can handle both incomplete and inconsistent information.

A *pre-bilattice* is a structure $\mathscr{B} = \langle B, \leq_t, \leq_k \rangle$, where B denotes a non-empty set and $\leq_t$ and $\leq_k$ are partial orderings on B. The ordering $\leq_k$ is thought of as ranking of "degree of information (or knowledge)". The bottom in $\leq_k$ is denoted by $\perp$ and the top by $\top$. If $x <_k y$, y gives us at least as much information as x (and possibly more).

The ordering $\leq_t$ is an ordering on the "degree of truth". The bottom in $\leq_t$ is denoted by false and the top by true. A bilattice can be obtained by adding certain assumptions for connections for two orderings.

One of the most well-known bilattices is the bilattice $FOUR$ as depicted as Fig. 2.3. The bilattice $FOUR$ can be interpreted as a combination of Belnap's lattices **A4** and **L4**.

The bilattice $FOUR$ can be seen as Belnap's lattice $FOUR$ with two kinds of orderings. Thus, we can think of the left-right direction as characterizing the ordering $\leq_t$: a move to the right is an increase in truth.

The meet operation $\wedge$ for $\leq_t$ is then characterized by: $x \wedge y$ is the rightmost thing that is of left both x and y. The join operation $\vee$ is dual to this. In a similar way, the up-down direction characterizes $\leq_k$: a move up is an increase in information. $x \otimes y$ is the uppermost thing below both x and y, and $\otimes$ is its dual.

To summarize; $\wedge$, $\vee$ and $\neg$ for the lattice operations that correspond to $\leq_t$. $\otimes$, $\oplus$, $-$ for the lattice operations that correspond to $\leq_k$. They are called *consensus*, *gullibility*, and *conflation*, respectively.

The symbols $\otimes$, $\oplus$ and $-$ are dual to $\wedge$, $\vee$ and $\neg$ in a sense that is aptly explained by referring to the lattices **L4** (associated with the truth order on **4**) and **A4** (associated with the information order on **4**). Originally, Ginsberg [11] called the structure $FOUR = \angle\mathbf{4}, \leq_t, \leq_k, \sim\rangle$ a bilattice as a prime example.

Fitting [9] gave a semantics for logic programming using bilattices. Kifer and Subrahmanian [16] interpreted Fitting's semantics within *generalized annotated logics* GAL.

A bilattice has a negation operation $\neg$ if there is a mapping $\neg$ that reverse $\leq_t$, leaves unchanged $\leq_k$ and $\neg\neg x = x$. Likewise, a bilattice has a conflation if there is a mapping $-$ that reverse $\leq_k$, leaves unchanged $\leq_t$. and $- - x = x$. If a bilattice has both operations, they commute if $-\neg x = \neg - x$ for all x.

In the bilattice $FOUR$, there is a negation operator under which $\neg t = f, \neg f = t$, and $\perp$ and $\top$ are left unchanged. There is also a conflation under which $-\perp = \top, -\top = \perp$ and t and f are left unchanged. And negation and conflation commute. In any bilattice, if a negation or conflation exists then the extreme elements $\perp, \top, f$ and t will behave as in $FOUR$.

Bilattice logics are theoretically elegant in that we can obtain several algebraic constructions, and are also suitable for reasoning about incomplete and inconsistent information. Arieli and Avron [4, 5] studied reasoning with bilattices. Thus, bilattice logics have many applications in AI as well as philosophy.

2.3.4 Relevance Logic

Relevance logic, also called *relevant logic*, is a family of logics based on the notion of relevance in conditionals. Historically, relevance logic was developed to avoid the *paradox of implications*; see Anderson and Belnap [2, 3].

Observe that some relevance logics are *paraconsistent* allowing contradiction, in the sense both A and $\neg A$ may be true. As relevance logic is closely related to many-valued logic, we give its survey.

Routley and Meyer proposed a basic relevant logic **B**, which is a minimal system having the so-called *Routley-Meyer semantics*l see Routley et al. [31].

The language of **B** contains logical symbols: $\sim$, &, $\vee$ and $\rightarrow$ (relevant implication). A Hilbert system for **B** is as follows:

Relevant Logic B

Axioms

(BA1) $A \rightarrow A$

(BA2) $(A \& B) \rightarrow A$

(BA3) $(A \& B) \rightarrow B$

(BA4) $((A \rightarrow B) \& (A \rightarrow C)) \rightarrow (A \rightarrow (B \& C))$

(BA5) $A \rightarrow (A \vee B)$

(BA6) $B \rightarrow (A \vee B)$

(BA7) $(A \rightarrow C) \& (B \rightarrow C)) \rightarrow ((A \vee B) \rightarrow C)$

(BA8) $(A \& (B \vee C)) \rightarrow (A \& B) \vee C)$

(BA9) $\sim\sim A \rightarrow A$

Rules of Inference

(BR1) $\vdash A, \vdash A \rightarrow B \Rightarrow \vdash B$

(BR2) $\vdash A, \vdash B \Rightarrow \vdash A\&B$

(BR3) $\vdash A \rightarrow B, \vdash C \rightarrow D \Rightarrow \vdash (B \rightarrow C) \rightarrow (A \rightarrow D)$

(BR4) $\vdash A \rightarrow \sim B \Rightarrow \vdash B \rightarrow \sim A$

Anderson and Belnap formalized a relevant logic **R** to realize a major motivation, in which they do not admit $A \rightarrow (B \rightarrow A)$. Later, various relevance logics have been proposed.

A Hilbert system for Anderson and Belnap's **R** is as follows:

Relevance Logic R

Axioms

(RA1) $A \rightarrow A$

(RA2) $(A \rightarrow B) \rightarrow ((C \rightarrow A) \rightarrow C \rightarrow B))$

(RA3) $(A \rightarrow (A \rightarrow B) \rightarrow (A \rightarrow B)$

(RA4) $(A \rightarrow (B \rightarrow C)) \rightarrow (B \rightarrow (A \rightarrow C)$

(RA5) $(A\&B) \rightarrow A$

(RA6) $(A\&B) \rightarrow B$

(RA7) $((A \rightarrow B)\&(A \rightarrow C)) \rightarrow (A \rightarrow (B\&C))$

(RA8) $A \rightarrow (A \vee B)$

(RA9) $B \rightarrow (A \vee B)$

(RA10) $((A \rightarrow C)\&(B \vee C)) \rightarrow ((A \vee B) \rightarrow C))$

(RA11) $(A\&(B \vee C)) \rightarrow ((A\&B) \vee C)$

(RA12) $(A \rightarrow\sim A) \rightarrow\sim A$

(RA13) $(A \rightarrow\sim B)) \rightarrow (B \rightarrow\sim A)$

(RA14) $\sim\sim A \rightarrow A$

Rules of Inference

(RR1) $\vdash A, \vdash A \rightarrow B \Rightarrow \vdash B$

(RR2) $\vdash A, \vdash B \Rightarrow \vdash A\&B$

Next, we give a Routley-Meyer semantics for **B**. A *model structure* is a tuple $\mathcal{M} = \langle K, N, R, *, v \rangle$, where K is a non-empty set of worlds, $N \subseteq K$, $R \subseteq K^3$ is a ternary relation on K, $*$ is a unary operation on K, and v is a valuation function from a set of worlds and a set of propositional variables $\mathcal{P}$ to $\{0, 1\}$.

There are some restrictions on $\mathcal{M}$. v satisfies the condition that $a \leq b$ and $v(a, p)$ imply $v(b, p) = 1$ for any $a, b \in K$ and any $p \in \mathcal{P}$. $a \leq b$ is a pre-order relation defined by $\exists x(x \in N$ and $Rxab)$. The operation $*$ satisfies the condition $a^{**} = a$.

For any propositional variable p, the truth condition $\models$ is defined: $a \models p$ iff $v(a, p) = 1$. Here, $a \models p$ reads "p is true at a". $\models$ can be extended for any formulas in the following way:

$$a \models \sim A \Leftrightarrow a^* \not\models A$$
$$a \models A \& B \Leftrightarrow a \models A \text{ and } a \models B$$
$$a \models A \vee B \Leftrightarrow a \models A \text{ or } a \models B$$
$$a \models A \rightarrow B \Leftrightarrow \forall bc \in K(Rabc \text{ and } b \models A \Rightarrow c \models B)$$

A formula A is *true* at a in $\mathscr{M}$ iff $a \models A$. A is *valid*, written $\models A$, iff A is true on all members of N in all model structures. It is well known that **B** is complete with respect to the above semantics using canonical models.

A model structure for **R** needs the following conditions.

$$R0aa$$
$$Rabc \Rightarrow Rbac$$
$$R^2(ab)cd \Rightarrow R^2a(bc)d$$
$$Raaa$$
$$a^{**} = a$$
$$Rabc \Rightarrow Rac^*b^*$$
$$Rabc \Rightarrow (a' \leq a \Rightarrow Ra'bc)$$

where R^2abcd is shorthand for $\exists x(Raxd$ and $Rxcd)$. The completeness theorem for the Routley-Meyer semantics can be proved for R; see [2, 3].

2.4 Basic Situation Calculus

In this section, we provide the basic situation calculus as defined by Reiter [30] and show how one can represent actions and their effects in this language.

The *situation calculus* is a second-order language specifically designed for representing dynamically changing worlds. All changes to the world are the result of named actions. A possible world history, which is simply a sequence of actions, is represented by a first-order term called a *situation*.

Generally, the values of relations and functions in a dynamic world will vary from one situation to the next. Relations whose truth values vary from situation to situation are called *relational fluents*. They are denoted by predicate symbols taking a situation term as their last argument.

Similarly, functions whose values vary from situation to situation are called functional fluents, and are denoted by function symbols taking a situation term as their last argument.

For example, in a world in which it is possible to paint objects, we might have a functional fluent $colour(x, s)$, denoting the colour of object x in that state of the world resulting from performing the action sequence s.

In a mobile robot environment, there might be a relational fluent $closeTo(r, x, s)$, meaning that in that state of the world reached by performing the action sequence s, the robot r will be close to the object x.

The principal intuition captured by our axioms is that situations are histories finite sequences of primitive actions and we provide a binary constructor $do(a, s)$ denoting the action sequence obtained from the history s by adding action a to it. Other intuitions are certainly possible about the nature of situations as discussed by McCarthy and Hayes [24] saw them as *snapshots* of a world.

The language $\mathscr{L}_{sitcalc}$ of the situation calculus is a second-order language with equality. It has three disjoint sorts: *action* for actions, *situation* for situations and a catch-all sort *object* for everything else depending on the domain of application.

Apart from the standard alphabet of logical symbols, we use logical symbols $\wedge$, $\neg$ and $\exists$ with the usual definitions of a full set of connectives and quantifiers. $\mathscr{L}_{sitcalc}$ has the following lexicon:

Countably infinitely many individual variable symbols of each sort. We shall use s and a, with subscripts and superscripts, for variables of sort *situation* and *action* respectively.

We normally use lower case roman letters other than a, s, with subscripts and superscripts for variables of sort *object.* In addition, because $\mathscr{L}_{sitcalc}$ is second-order, its alphabet includes countably infinitely many predicate variables of all arities.

Two function symbols of sort *situation* are

1. A constant symbol S_0 denoting the initial situation.
2. A binary function symbol $do : action \times situation \rightarrow situation.$
   The intended interpretation is that $do(a, s)$ denotes the successor situation resulting from performing action a in situation s.

A binary predicate symbol $\sqsubset$: $situation \times situation$, defining an ordering relation on situations. The intended interpretation of situations is as action histories, in which case $s \sqsubset s'$ means that s is a proper subhistory of s'.

A binary predicate symbol $Poss : action \times situation$. The intended interpretation of $Poss(a, s)$ is that it is possible to perform the action a in situation s.

For each $n \geq 0$, countably infinitely many predicate symbols with arity n, and sorts $(action \cup object)^n$. These are used to denote situation independent relations like $human(John)$, $primeNumber(n)$, $movingAction(run(person, loc1, loc2))$, etc.

For each $n \geq 0$, countably infinitely many function symbols of sort $(action \cup object)^n \rightarrow object$. These are used to denote situation independent functions like $sqrt(x)$, $height(MtEverest)$, $agent(run(person, loc1, loc2))$, etc.

For each $n \geq 0$, a finite or countably infinite number of function symbols of sort $(action \cup object)^n \rightarrow action$. These are called *action functions* and are used to denote actions like $pickup(x)$, $move(A, B)$, etc.

In most applications, there will be just finitely many action functions, but we allow the possibility of an infinite number of them. For each $n \geq 0$, a finite or countably infinite number of predicate symbols with arity $n + 1$ and sorts $(action \cup object)^n \times situation$. These predicate symbols are called *relational fluents*.

In most applications, there will be just finitely many relational fluents, but we do not preclude the possibility of an infinite number of them. These are used to denote situation dependent relations like $ontable(x, s)$, $husband(Mary, John, s)$, etc. Notice that relational fluents take just one argument of sort *situation*, and this is always its last argument.

For each $n \geq 0$, a finite or countably infinite number of function symbols of sort $(action \cup object)^n \times situation \rightarrow action \cup object$. These function symbols are called functional fluents.

In most applications there will be just finitely many functional fluents, but we do not preclude the possibility of an infinite number of them. These are used to denote situation dependent functions like $age(Mary, s)$, $prime(Minister, Italy, s)$, etc. Notice that functional fluents take just one argument of sort situation, and this is always its last argument.

2.4.1 Foundational Axioms for Situations

We now focus on the domain of situations. The primary intuition about situations that we wish to capture axiomatically is that they are finite sequences of actions. We also want to say that a certain sequence of actions precedes another.

The four axioms we are about to present capture these two properties of situations:

$$do(a_1, s_1) = do(a_2, s_2) \rightarrow a_1 = a_2 \wedge s_1 = s_2 \tag{2.1a}$$

$$(\forall P).P(S_0) \wedge (\forall a, s)[P(s) \rightarrow P(do(a, s))] \rightarrow (\forall s)P(s) \tag{2.1b}$$

Compare these to the first two axioms for the natural numbers.

Axiom (2.1b) is a second-order induction axiom, and has the effect of limiting the sort situation to the smallest set containing S_o, and closed under the application of the function do to an action and a situation.

Any model of these axioms will have as its domain of situations the smallest set S satisfying:

1. $\sigma_0 \in S$, where a_0 is the interpretation of S_0 in the model.
2. If $\sigma \in \mathscr{S}$, and $A \in \mathscr{A}$, then $do(A, \sigma) \in \mathscr{S}$, where $\mathscr{A}$ is the domain of actions in the model.

Notice that axiom (2.1) is a unique name axiom for situations. This, together with the induction axiom, implies that two situations will be the same if they result from the same sequence of actions applied to the initial situation.

Two situations S_1 and S_2 may be different, yet assign the same truth values to all fluents. So a situation in the situation calculus must not be identified with the set of fluents that hold in that situation, i.e. with a state.

The proper way to understand a situation is as history, namely, a finite sequence of actions; two situations are equal iff they denote identical histories.

This is the major reason for using the terminology *situation* instead of *state*; the latter carries with it the connotation of a *snapshot* of the world.

In our formulation of the situation calculus, situations are not snapshots, they are finite sequences of actions. While states can repeat themselves–the same snapshot of the world can happen twice–situations cannot.

There are two more axioms, designed to capture the concept of a subhistory:

$$\neg s \sqsubset S_0, \tag{2.2a}$$

$$s \sqsubset do(a, s') \equiv s \sqsubseteq s', \tag{2.2b}$$

where $s \sqsubseteq s'$ is an abbreviation for $s \sqsubset s' \vee s = s'$.

Here, the relation $\sqsubset$ provides an ordering on situations; $s \sqsubset s'$ means that the action sequence s' can be obtained from the sequences by adding one or more actions to the front of s. These axioms also have their analogues in the last two axioms of the preceding fragment of number theory.

2.4.2 Domain Axioms and Basic Theories of Actions

Our concern here is with axiomatizations for actions and their effects that have a particular syntactic form. These are called basic action theories, and we next describe these.

Definition 2.4 (The Uniform Formulas) Let σ be a term of sort situation. The terms of $\mathcal{L}_{sitcalc}$ uniform in σ are the smallest set of terms such that:

1. Any term that does not mention a term of sort *situation* is uniform in σ.
2. σ is uniform in σ.
3. If g is an n-ary function symbol other than do and S_0, and $t_1, \ldots, t_n$ are terms uniform in σ whose sorts are appropriate for g then $g(t_1, \ldots, t_n)$ is a term uniform in σ.

The formulas of $\mathcal{L}_{sitcalc}$ uniform in σ are the smallest set of formulas such that:

1. If t_1 and t_2 are terms of the same sort object or action, and if they are both uniform in σ, then $t_1 = t_2$ is a formula uniform in σ.
2. When P is an n-ary predicate symbol of $\mathcal{L}_{sitcalc}$ other than $Poss$ and $\sqsubset$, and $t_1, \ldots, t_n$ are terms uniform in σ whose sorts are appropriate for P, then $P(t_1, \ldots, t_n)$ is a formula uniform in σ.
3. Whenever U_1, U_2 are formulas uniform in σ so are $\neg U_1$, $U_1 \wedge U_2$ and $(\exists v)U_1$ provided v is an individual variable, and it is not of sort *situation*.

Thus, a formula of $\mathcal{L}_{sitcalc}$ is uniform in σ iff it is first order, it does not mention the predicates $Poss$ or $\sqsubset$ it does not quantify over variables of sort *situation*, it does not mention equality on situations, and whenever it mentions a term of sort *situation* in the situation argument position of a fluent, then that term is σ.

Definition 2.5 (Action Precondition Axiom) An action precondition axiom of Lsitcalc is a sentence of the form:

$$Poss(A(x_1, \ldots, x_n), s) \equiv \Pi_A(x_1, \ldots, x_n, s),$$

where A is an $n-$ary action function symbol, and $\Pi_A(x_1, \ldots, x_n, s)$ is a formula that is uniform in s and whose free variables are among $x_1, \ldots, x_n, s$. For example, in a blocks world, we might typically have;

$$Poss(pickup(x), s) \equiv (\forall y)\neg holding(y, s) \wedge \neg heavy(x, s).$$

The uniformity requirement on Π_A ensures that the preconditions for the executability of the action $A(x_1, \ldots, x_n)$ are determined only by the current situation s, not by any other situation.

Definition 2.6 (Successor State Axiom)

1. A successor state axiom for an $(n + 1)$-ary relational fluent F is a sentence of $\mathscr{L}_{sitcalc}$ of the form:

   $$F(x_1, \ldots, x_n, do(a, s)) \equiv \Phi_F(x_1, \ldots, x_n, a, s),$$

   where $\Phi_F(x_1, \ldots, x_n, a, s)$ is a formula uniform in s, all of whose free variables are among $a, s, x_1, \ldots, x_n$.

   An example of such an axiom is:

   $$\begin{aligned}&broken(x, do(a, s)) \equiv\\ &(\exists r)\{a = drop(r, x) \wedge fragile(x, s)\} \vee (\exists b)\{a = explode(b) \wedge nexto(b, x, s)\} \vee\\ &broken(x, s) \wedge \neg(\exists r)a = repair(r, x).\end{aligned}$$

   As for action precondition axioms, the uniformity of Φ_F guarantees that the truth value of $F(x_1, \ldots, x_n, do(a, s))$ in the successor situation $do(a, s)$ is determined entirely by the current situation s, and not by any other situation. In systems and control theory, this is often called the *Markov property*.

2. A successor state axiom for an $(n + 1)$-ary functional fluent f is a sentence of $\mathscr{L}_{sitcalc}$ of the form:

   $$f(x_1, \ldots, x_n, do(a, s)) = y \equiv \phi_f(x_1, \ldots, x_n, y, a, s),$$

   where $\phi_f(x_1, x_1, \ldots, x_n, y, a, s)$ is a formula uniform in s, all of whose free variables are among $x_1, \ldots, x_n, y, a, s)$.

   A blocks world example is:

   $$\begin{aligned}&height(x, do(a, s)) = y \equiv a = moveToTable(x) \wedge y = 1 \vee\\ &(\exists z, h)(a = move(x, z) \wedge height(z, s) = h + y = h + 1) \vee\\ &height(x, s) = y \wedge a \neq moveToTable(x) \wedge \neg(\exists z)a) = move(x, z).\end{aligned}$$

   As for relational fluents, the uniformity of ϕ_f in the successor state axioms for functional fluents guarantees the Markov property.

   Thus, the value of a functional fluent in a successor situation is determined entirely by properties of the current situation, and not by any other situations.

Basic Action Theories: Henceforth, we shall consider theories $\mathcal{D}$ of $\mathcal{L}_{sitcalc}$ of the following forms:

$\mathcal{D} = \Sigma \cup \mathcal{D}_{ss} \cup \mathcal{D}_{ap} \cup \mathcal{D}_{una} \cup \mathcal{D}_{S_0}$

where,

1. Σ are the foundational axioms for situations.
2. $\mathcal{D}_{ss}$ is a set of successor state axioms for functional and relational fluents, one for each such fluent of the language $\mathcal{L}_{sitcalc}$
3. $\mathcal{D}_{ap}$ is a set of action precondition axioms, one for each action function symbol of $\mathcal{L}_{sitcalc}$
4. $\mathcal{D}_{una}$ is the set of unique names axioms for all action function symbols of $\mathcal{L}_{sitcalc}$
5. $\mathcal{D}_{S_0}$ is a set of first order sentences that are uniform in S_0. Thus, no sentence of $\mathcal{D}_{S_0}$ quantifies over situations, or mentions $Poss$, $\sqsubset$ or the function symbol do, so that S_0 is the only term of sort situation mentioned by these sentences, $\mathcal{D}_{S_0}$ will function as the initial theory of the world (i.e. the one we start off with, before any actions have been "executed")

   Often, we shall call $\mathcal{D}_{S_0}$ the *initial database*. The initial database may (and often will) contain sentences mentioning no situation term at all, for example, unique names axioms for individuals, like $John \neq Mary$, or "timeless" facts like $isMountain(MtEverest)$, or $dog(x) \rightarrow mammal(x)$.

Definition 2.7 A *basic action theory* is any collection of axioms $\mathcal{D}$ of the above form that also satisfies the following *functional fluent consistency property*.

Whenever f is a functional fluent whose successor state axiom in $\mathcal{D}_{ss}$ is

$$f(\mathbf{x}, do(a, s)) = y \equiv \phi_f(\mathbf{x}, y, a, s),$$

then

$$\mathcal{D}_{una} \cup \mathcal{D}_{S_0} \models (\forall a, s).(\forall \mathbf{x}).(\exists y)\phi_f(\mathbf{x}, y, a, s) \wedge$$
$$[(\forall y, y').\phi_f(\mathbf{x}, y, a, s) \wedge \phi_f(\mathbf{x}, y', a, s) \rightarrow y = y'].$$

Regression is perhaps the single most important theorem-proving mechanism for the situation calculus; it provides a systematic way to establish that a basic action theory entails a so-called regressable sentence.

Reiter [30] proved decidability of a fragment of the situation calculus, so-called *propositional fluents*, basic action theories and regressable formulas. But in this book, we do not treat Regression for decidability.

As for the Frame Problem, we will give a survey and discussion in the context of granular reasoning in Chap. 5.

2.4.3 The Frame Problem

Now, we consider axiomatizing actions in the Situation Calculus; see Reiter [30]. The first observation one can make about actions is that they have preconditions: requirements that must be satisfied whenever they can be executed in the current situation.

We introduce a predicate symbol $Poss$; $Poss(a, s)$ means that it is possible to perform the action a in that state of the world resulting from performing the sequence of actions.

Here are some examples:

- If it is possible for a robot r to pick up an object x in situation s, then the robot is not holding any object, it is next to x, and x is not heavy:

  $$Poss(pickup(r, x), s) \rightarrow [(\forall z)\neg holding(r, z, s)] \wedge heavy(x) \wedge nextTo(r, x, s).$$

- Whenever it is possible for a robot to repair an object, then the object must be broken, and there must be glue available:

  $$Poss(repair(r, x), s) \rightarrow hasGlue(r, s) \wedge broken(x, s).$$

The next feature of dynamic worlds that must be described are the causal laws—how actions affect the values of fluents. These are specified by so-called effect axioms. The following are some examples:

- The effect on the relational fluent broken of a robot dropping a fragile object:

  $$fragile(x, s) \rightarrow broken(xdo(drop(r, x), s)).$$

This is the situation calculus way of saying that dropping a fragile object causes it to become broken; in the current situation s, if x is fragile, then in that successor situation $do(drop(r, x), s)$ resulting from performing the action $drop(r, x)$ in s, x will be broken.

- A robot repairing an object causes it not to be broken:

  $$\neg broken(x, do(repair(r, x), s)).$$

- Painting an object with colour c:

  $$colour(x, do(paint(x, c), s)) = c.$$

The Qualification Problem for Actions

With only the above axioms, nothing interesting can be proved about when an action is possible. For example, here are some preconditions for the action pickup:

$$Poss(pickup(r, x), s) \rightarrow [(\forall z)\neg holding(r, z, s)] \wedge \neg(x) \wedge nextTo(r, x, s).$$

The reason nothing interesting follows from this is clear; we can never infer when a pickup is possible. We can try reversing the implication:

$$[(\forall z)\neg holding(r, z, s)] \wedge \neg heavy(x) \wedge nextTo(r, x, s) \rightarrow Poss(pickup(r, x), s).$$

Now we can indeed infer when a pickup is possible, but unfortunately, this sentence is false. We also need, in the antecedent of the implication:

$$\neg gluedToFloor(x, s) \wedge \neg armsTied(r, s) \wedge \neg hitByTenTonTruck(r, s) \wedge \cdots$$

i.e, we need to specify all the qualifications that must be true in order for a pickup to be possible! For the sake of argument, imagine succeeding in enumerating all the qualifications for pickup.

Suppose the only facts known to us about a particular robot R, object A, and situation S are:

$$[(\forall z)\neg holding(R, z, S)] \wedge \neg heavy(a) \wedge nextTo(R, A, S).$$

We still cannot infer $Poss(pickup(R, A), S)$ because we are not given that the above qualifications are true! Intuitively, here is what we want: When given only that the "important" qualifications are true:

$$[(\forall z)\neg holding(R, z, S)] \wedge \neg heavy(a) \wedge nextTo(R, A, S).$$

and if we do not know that any of the "minor" qualifications—$\neg gluedToFloor(A, S)$, $\neg hitBvTenTonTruck(R, S)$—are true, infer $Poss(pickup(R, A), S)$.

But if we happen pen to know that any one of the minor qualifications is false, this will block the inference of $Poss(pickup(R, A), S)$.

Historically, this has been seen to be a problem peculiar to reasoning about actions, but this is not really the case.

Consider the following fact about $birds$, which has nothing to do with reasoning about actions:

$$bird(x) \wedge \neg penguin(x) \wedge \neg ostrich(x) \wedge \neg pekingDuck(x) \wedge \cdots \rightarrow flies(x).$$

But given only the fact $bird(Tweety)$, we want intuitively to infer $flies(Tweety)$. Formally, this is the same problem as action qualifications:

- The important qualification is $bird(x)$.
- The minor qualifications are: $\neg penguin(x)$, $\neg ostrich(x)$, $\cdots$

This is the classical example of the need for non-monotonic reasoning in artificial intelligence (AI).

For the moment, it is sufficient to recognize that the qualification problem for actions is an instance of a much more general problem, and that there is no obvious way to address it.

We shall adopt the following (admittedly idealized) approach: Assume that for each action A(x), there is an axiom of the form

$$Poss(A((x), s) \equiv \Pi_A(\mathbf{x}, s),$$

where $Poss(A((x), s)$ is a first-order formula with free variables x, s that does not mention the function symbol do. We shall call these action precondition axioms. An example is:

$$Poss(pickup(r, x), s) \equiv [(\forall z)\neg holding(r, z, s)] \wedge \neg heavy(x) \wedge nextTo(r, x, s).$$

These axioms say that we choose to ignore all the "minor" qualifications, in favour of necessary and sufficient conditions defining when an action can be performed.

2.4.4 Frame Axioms

There is another well known problem associated with axiomatizing dynamic worlds; axioms other than effect axioms are required. These are called *frame axioms*, and they specify the action invariants of the domain, i.e, those fluents unaffected by the performance of an action.

For example, the following is a positive frame axiom, declaring that the action of robot r' painting object x' with colour c has no effect on robot r holding object x:

$$holding(r, x, s) \rightarrow holding(r, x, do(print(r', x'), s)).$$

Here is a negative frame axiom for not breaking things:

$$\neg broken(x, s) \wedge [x \neq y \vee \neg fragile(x, s)] \rightarrow \neg broken(x, do(drop(r, y), s)).$$

Notice that these frame axioms are truths about the world, and therefore must be included in any formal description of the dynamics of the world.

The problem is that there will be a vast number of such axioms because only relatively few actions will affect the value of a given fluent. All other actions leave the fluent invariant.

For example, an object's colour remains unchanged after picking something up, opening a door, turning on a light, electing a new prime minister of Canada, etc.

Since, empirically in the real world, most actions have no effect on a given fluent, we can expect of the order of $2 \times \mathscr{A} \times \mathscr{F}$ frame axioms, where $\mathscr{A}$ is the number of actions, and $\mathscr{F}$ the number of fluents.

These observations lead to what is called *The Frame Problem*:

1. The axiomatizer must think of, and write down, all these quadratically many frame axioms. In a setting with 100 actions and 100 fluents, this involves roughly 20, 000 frame axioms.
2. The implementation must somehow reason efficiently in the presence of so many axioms.

Suppose the person responsible for axiomatizing an application domain has specified all the causal laws for that domain.

More precisely, she has succeeded in writing down all the effect axioms, i.e. for each relational fluent $\mathscr{F}$ and each action $\mathscr{A}$ that causes F's truth value to change, axioms of the form:

$$R(\mathbf{x}, s) \rightarrow (\neg)F(\mathbf{x}, do(A, s)),$$

and for each functional fluent f and each action A that can cause f's value to change, axioms of the form:

$$R(\mathbf{x}, y, s) \rightarrow f(\mathbf{x}, do(A, s)) = y.$$

Here, R is a first-order formula specifying the contextual conditions under which the action A will have its specified effect on F and f.

There are no restrictions on R, except that it must refer only to the current situation s. Later, we shall be more precise about the syntactic form of these effect axioms.

A solution to the Frame Problem is therefore a systematic procedure for generating, from these effect axioms, all the frame axioms. And, if possible, we also want a parsimonious representation for these frame axioms (because in their simplest form, there are too many of them).

We finished the discussion on the Frame Problem. In later chapters, we will discuss its philosophical significance, and elaborate on it in connection with granular reasoning, which is one of the objectives of the present book.

References

1. Akama, S.: Non-classical logics and intelligent systems. In: Nakamatsu, K., Jain, L. (eds.) The Handbook on Reasoning-based Intelligent Systems, pp. 189–205. World Scientific, Singapore (2013)
2. Anderson, A., Belnap, N.: Entailment: The Logic of Relevance and Necessity I. Princeton University Press, Princeton (1976)
3. Anderson, A., Belnap, N., Dunn, J.: Entailment: The Logic of Relevance and Necessity II. Princeton University Press, Princeton (1992)
4. Arieli, O., Avron, A.: Reasoning with logical bilattices. J. Logic Lang. Inform. **5**, 25–63 (1996)
5. Arieli, O., Avron, A.: The value of the four values. Artif. Intell. **102**, 97–141 (1998)
6. Belnap, N.D.: A useful four-valued logic. In: Dunn, J.M., Epstein, G. (eds.) Modern Uses of Multi-Valued Logic, pp. 8–37. Reidel, Dordrecht (1977)

7. Belnap, N.D.: How a computer should think. In: Ryle, G. (ed.) Contemporary Aspects of Philosophy, pp. 30–55. Oriel Press (1977)
8. Fitting, M.: A Kripke/Kleene semantics for logic programs. J. Logic Program. **2**, 295–312 (1985)
9. Fitting, M.: Bilattices and the semantics of logic programming. J. Logic Program. **11**, 91–116 (1991)
10. Fitting, M.: A theory of truth that prefers falsehood. J. Philos. Logic **26**, 477–500 (1997)
11. Ginsberg, M.: Multivalued logics: a uniform approach to reasoning in AI. Comput. Intell. **4**, 256–316 (1988)
12. Hajek, P.: Metamathematics of Fuzzy Logic. Kluwer, Dordrecht (1998)
13. Hintikka, J.: Knowledge and Belief: An Introduction to the Logic of the Two Notions. Cornell University Press, Ithaca (1962)
14. Hughes, G., Cresswell, M.: An Introduction to Modal Logic. Methuen, London (1968)
15. Hughes, G., Cresswell, M.: A New Introduction to Modal Logic. Routledge, London (1996)
16. Kifer, M., Subrahmanian, V.S.: On the expressive power of annotated logic programs. In: Proc. of the 1989 North American Conference on Logic Programming, pp. 1069–1089 (1989)
17. Kleene, S.: Introduction to Metamathematics. North-Holland, Amsterdam (1952)
18. Kripke, S.: A complete theorem in modal logic. J. Symbol. Logic **24**, 1–24 (1959)
19. Kripke, S.: Semantical considerations on modal logic. Acta Philos. Fennica **16**, 83–94 (1963)
20. Kripke, S.: Semantical analysis of modal logic I. Z. für Math. Logik Grundlagen Math. **9**, 67–96 (1963)
21. Kunen, K.: Negation in logic programming. J. Logic Program. **4**, 289–308 (1987)
22. Łukasiewicz, J.: On 3-valued logic, 1920. In: McCall, S. (ed.) Polish Logic, pp. 16–18. Oxford University Press, Oxford (1967)
23. Łukasiewicz, J.: Many-valued systems of propositional logic, 1930. In: McCall, S. (ed.) Polish Logic. Oxford University Press, Oxford (1967)
24. McCarthy, J., Hayes, P.: Some philosophical problems from the standpoint of artificial intelligence. In: Meltzer, B., Michie, D. (eds.) Machine Intelligence, vol. 4, pp. 463–502. Edinburgh University Press, Edinburgh (1969)
25. Pawlak, Z.: Rough sets. Int. J. Comput. Inf. Sci. **11**, 341–356 (1982)
26. Pawlak, Z.: Rough Sets: Theoretical Aspects of Reasoning about Data. Kluwer, Dordrecht (1991)
27. Priest, G.: An Introduction to Non-Classical Logic. Cambridge University Press, Cambridge (2001)
28. Priest, G.: An Introduction to Non-Classical Logic, 2nd edn. Cambridge University Press, Cambridge (2008)
29. Prior, A.: Past, Present and Future. Clarendon Press, Oxford (1967)
30. Reiter, R.: Knowledge in Action: Logical Foundations for Specifying and Implementing Dynamical Systems. MIT Press, Cambridge, Mass (2001)
31. Routley, R., Plumwood, V., Meyer, R.K., Brady, R.: Relevant Logics and Their Rivals, vol. 1. Ridgeview, Atascadero (1982)
32. Shen, Y., Wang, F.: Variable precision rough set model over two universes and its properties. Soft Comput. **15**, 557–567 (2011)
33. Zadeh, L.: Fuzzy sets. Inf. Control **8**, 338–353 (1965)
34. Ziarko, W.: Variable precision rough set model. J. Comput. Syst. Sci. **46**, 39–59 (1993)

Chapter 3
Deduction Systems Based on Rough Sets

Abstract In Chap. 3, we present the consequence relation and Gentzen type sequent calculus for many-valued logics, and also describe the partial semantics interpreted with rough sets.

3.1 Introduction

This chapter is based on Nakayama et al. [17, 18]. *Rough set theory* has been extensively used both as a mathematical foundation of granularity and vagueness in information systems and in a large number of applications. However, the decision logic for rough sets is based on classical bivalent logic; therefore, it would be desirable to develop decision logic for handling uncertain, ambiguous and inconsistent objects.

In this chapter, a deduction system based on partial semantics is proposed for decision logic. We propose Belnap's four-valued semantics as the basis for three-valued and four-valued logics to extend the deduction of decision logic since the boundary region of rough sets is interpreted as both a non-deterministic and inconsistent state.

We also introduce the consequence relations to serve as an intermediary between rough sets and many-valued semantics. Hence, consequence relations based on partial semantics for decision logic are defined, and axiomatization by Gentzen-type sequent calculi is obtained.

Furthermore, we extend the sequent calculi with a weak implication to hold for a deduction theorem and also show a soundness and completeness theorem for the four-valued logic for decision logic.

Pawlak introduced the theory of rough sets for handling rough (coarse) information [19, 20]. Rough set theory is now used as a mathematical foundation of granularity and vagueness in information systems and is applied to a variety of problems. In applying rough set theory, decision logic was proposed for interpreting information extracted from data tables.

However, decision logic adopts the classical two-valued logic semantics. It is known that classical logic is not adequate for reasoning with indefinite and inconsistent information. Moreover, the paradoxes of the material implications of classical logic are counterintuitive.

S. Akama et al., *Epistemic Situation Calculus Based on Granular Computing*,
Intelligent Systems Reference Library 239,
https://doi.org/10.1007/978-3-031-28551-6_3

Rough set theory can handle the concept of approximation by the indiscernibility relation, which is a central concept in rough set theory. It is an equivalence relation, where all identical objects of sets are considered elementary. Rough set theory is concerned with the lower and upper approximations of object sets.

These approximations divide sets into three regions, namely, the positive, negative, and boundary regions. Thus, Pawlak rough sets have often been studied in a three-valued logic framework because the third value is thought to correspond to the boundary region of rough sets [4, 7].

On the contrary, in this chapter, we propose that the interpretation of the boundary region is based on four-valued semantics rather than three-valued since the boundary region can be interpreted as both undefined and overdefined.

For example, a knowledge base K of a rough set can be seen as a theory *KB* whose underlying logic is L. *KB* is called *inconsistent* when it contains theorems of the form A and $\sim A$ (the negation of A).

If *KB* is not inconsistent, it is called *consistent*. Our approach for a rough set proposes useful theory to handle such inconsistent information without system failure. In this study, non-deterministic features are considered the characteristic of partial semantics. Undetermined objects in the boundary region of rough sets have two interpretations of both undefinedness and inconsistency.

The formalization of both three-valued and four-valued logics is carried out using a consequence relation based on partial semantics. The basic logic for decision logic is assumed to be many-valued, in particular, three-valued or four-valued and some of its alternatives [24].

If such many-valued logics are used as a basic deduction system for decision logic, it can be enhanced to a more useful method for data analysis and information processing. The decision logic of rough set theory will be axiomatized using Gentzen sequent calculi and a four-valued semantic relation as basic theory.

To introduce many-valued logic to decision logic, consequence relations based on partial interpretation are investigated, and the sequent calculi of many-valued logic based on them are constructed. Subsequently, many-valued logics with weak implication are considered for the deduction system of decision logic.

The deductive system of decision logic has been studied from the granule computing perspective, and in [11], an extension of decision logic was proposed for handling uncertain data tables by fuzzy and probabilistic methods. In [13], a natural deduction system based on classical logic was proposed for decision logic in granule computing.

In [4], the sequent calculi of the Kleene and Łukasiewicz three-valued logics were proposed for rough set theory based on non-deterministic matrices for semantic interpretation. The Gentzen-type axiomatization of three-valued logics based on partial semantics for decision logic is proposed in [17, 18]. The reasoning for rough sets is comprehensively studied in [1].

This chapter is organized as follows. In Sect. 3.2, we briefly review rough sets, the decision table, and decision logic. In Sect. 3.3, Belnap's four-valued semantics is introduced as the basis of the semantics interpretation presented in the chapter.

In Sect. 3.4, we present a partial semantics model for rough sets and decision logic based on four-valued semantics, and some characteristics are presented.

In Sect. 3.5, an axiomatization using Gentzen sequent calculus is presented according to a consequence relation based on the previously discussed partial semantics. In Sect. 3.6, we discuss the extension of sequent calculi for many-valued logics with weak negation and implication to enable a deduction theorem. In Sect. 3.7, the soundness and completeness theorems are shown for a four-valued sequent calculus GC4. Finally, in Sect. 3.8, a summary of the study and possible directions for future work are provided.

3.2 Rough Sets and Decision Logic

Rough set theory, proposed by Pawlak [19], provides a theoretical basis of sets based on approximation concepts. A rough set can be seen as an approximation of a set. It is denoted by a pair of sets called the lower and upper approximations of the set. Rough sets are used for imprecise data handling.

For the upper and lower approximations, any subset X of U can be in any of three states according to the membership relation of the objects in U. If the positive and negative regions on a rough set are considered to correspond to the truth-value of a logical form, then the boundary region corresponds to ambiguity in deciding truth or falsity. Thus, it is natural to adopt a three-valued logic.

Rough set theory is outlined below. Let U be a non-empty finite set called a universe of objects. If R is an equivalence relation on U, then U/R denotes the family of all equivalence classes of R, and the pair (U, R) is called a Pawlak approximation space. A knowledge base K is defined as follows:

Definition 3.1 A knowledge base K is a pair $K = (U, R)$, where U is a universe of objects, and $\mathbf{R}$ is a set of equivalence relations on the objects in U.

Definition 3.2 Let $R \in \mathbf{R}$ be an equivalence relation of the knowledge base $K = (U, R)$ and X any subset of U. Then, the lower and upper approximations of X for R are defined as follows:

$$\underline{R}X = \bigcup\{Y \in U/R \mid Y \subseteq X\} = \{x \in U \mid [x]_{\mathrm{R}} \subseteq X\},$$
$$\overline{R}X = \bigcup\{Y \in U/R \mid Y \cap X \neq 0\} = \{x \in U \mid [x]_{\mathrm{R}} \cap X \neq \emptyset\}.$$

Definition 3.3 If $K = (U, R)$, $R \in \mathbf{R}$, and $X \subseteq U$, then the R-positive, R-negative, and R-boundary regions of X with respect to R are defined respectively as follows:

$$POS_R(X) = \underline{R}X,$$
$$NEG_R(X) = U - \overline{R}X,$$
$$BN_R(X) = \overline{R}X - \underline{R}X.$$

Objects included in an R-boundary are interpreted as the truth-value gap or glut. The semantic interpretation for rough sets is defined later.

3.2.1 Decision Tables

Decision tables can be seen as a special important class of knowledge representation systems and can be used for applications. Let $K = (U, A)$ be a knowledge representation system and $C, D \subset A$ be two subsets of attributes called condition and decision attributes, respectively.

A KR-system with a distinguished condition and decision attributes is called a decision table, denoted $T = (U, A, V, s)$ or in short DC, where U is a finite and nonempty set of objects, A is a finite and nonempty set of attributes, V is a nonempty set of values for $a \in A$, and s is an information function that assigns a value $U \times s_x : A \rightarrow V$ (for simplicity, the subscript x will be omitted), where $\forall x \in U$, and $\forall a \in C \cup D \subset A$.

Equivalence classes of the relations $IND(C)$ and $IND(D)$, a subset of A, are called *condition* and *decision classes*, respectively.

With every $x \in U$, we associate a function $dx : A \rightarrow V$, such that $d_x(a) = a(x)$ for every $a \in C \cup D$; the function d_x is called a *decision rule* (in T), and x is referred to as a label of the decision rule d_x.

The decision rule d_x is *consistent* (in T) if for every $y \neq x$, $d_x|C = d_y|C$ implies $d_x|D = d_y|D$; otherwise the decision rule is *inconsistent*.

A decision table is *consistent* if all of its decision rules are consistent; otherwise the decision table is *inconsistent*. Consistency (inconsistency) sometimes may be interpreted as determinism (non-determinism).

Proposition 3.1 *A decision table $T = (U, A, V, s)$ is consistent iff $C \Rightarrow D$, where C and D are condition and decision attributes.*

From Proposition 3.1, it follows that the practical method of checking the consistency of a decision table is by simply computing the degree of dependency between the condition and decision attributes. If the degree of dependency equals 1, then we conclude that the table is consistent; otherwise, it is inconsistent.

Consider Table 3.1 from Pawlak [20].

Table 3.1 Decision table

U	A	B	C	D	E
1	1	0	2	2	0
2	0	1	1	1	2
3	2	0	0	2	2
4	1	0	2	2	0
5	1	0	2	0	1
6	2	2	0	1	1
7	2	1	1	1	2
8	0	1	1	0	1

Assume that a, b, and c are condition attributes and d and e are decision attributes. In this table, for instance, decision rule 1 is inconsistent, whereas decision rule 3 is consistent. Decision rules 1 and 5 have the same condition, but their decisions are different.

3.2.2 Decision Logic

A decision logic language (DL-language) L is now introduced cf. [19]. The set of attribute constants is defined as $a \in A$, and the set of attribute value constants is $V = \bigcup V_a$. The propositional variables are φ and ψ, and the propositional connectives are $\perp$, $\sim$, $\wedge$, $\vee$, $\rightarrow$ and $\equiv$.

Definition 3.4 The set of formulas of the decision logic language (DL-language) L is the smallest set satisfying the following conditions:

1. (a, v), or in short a_v, is an atomic formula of L.
2. If φ *and* ψ are formulas of the DL-language, then $\sim \varphi$, $\varphi \wedge \psi$, $\varphi \vee \psi$, $\varphi \rightarrow \psi$, and $\varphi \equiv \psi$ are formulas.

The interpretation of the DL-language L is performed using the universe U in $S = (U, A)$ of the Knowledge Representation System (*KR-system*) and the assignment function, mapping from U to objects of formulas. Formulas of the DL-language are interpreted as subsets of objects consisting of a value v and an attribute a.

Atomic formulas (a, v) describe objects that have a value v for the attribute a. Attribute a is a function from U to V, defined by $a(x) = s_x(a)$, where $x \in U$, and $s_x(a) \in V$. If let $s_x(a) = v$, then a can be viewed as a binary relation on U, such that for $\langle x, v\rangle \in U \times U$, $\langle a, v\rangle \in a$ if and only if $a(x) = v$.

In this case, the atomic formula (a, v) can be denoted by $a(x, v)$, where x is a variable, and v is taken as a constant; they are all terms in U. Thus, (a, v) can be viewed as formula $a(x, v)$ which is an atomic formula.

The semantics for *DL* is given by a model. For DL, the model is the KR-system $S = (U, A)$, which describes the meaning of symbols of predicates (a, v) in U, and if we properly interpret the formulas in the model, then each formula becomes a meaningful sentence, expressing the properties of some objects. An object $x \in U$ satisfies a formula φ in $S = (U, A)$, denoted $x \models_S \varphi$ or in short $x \models \varphi$, iff the following conditions are satisfied:

Definition 3.5 The semantic relations of a DL-language are defined as follows:

$x \models_S a(x, v)$ iff $a(x) = v$,
$x \models_S \sim \varphi$ iff $x \not\models_S \varphi$,
$x \models_S \varphi \vee \psi$ iff $x \models_S \varphi$ *or* $x \models_S \psi$,
$x \models_S \varphi \wedge \psi$ iff $x \models_S \varphi$ *and* $S \models_S \psi$,
$x \models_S \varphi \rightarrow \psi$ iff $x \models_S l \sim \varphi \vee \psi$,
$x \models_S \varphi \equiv \psi$ iff $x \models_S \varphi \rightarrow \psi$ and $s \models_S \psi \rightarrow \varphi$.

If φ is a formula, then the set $|\varphi|_S$ defined as follows:

$$|\varphi|_S = \{x \in U \mid x \models_S \varphi\}$$

and will be called the meaning of the formula φ in S. The following properties are obvious:

Proposition 3.2 *The meaning of an arbitrary formula satisfies the following:*
$|\neg\varphi|_S = U - |\varphi|_S,$
$|\varphi \vee \psi|_S = |\varphi|_S \cup |\psi|_S,$
$|\varphi \wedge \psi|_S = |\varphi|_S \cap |\psi|_S,$
$|\varphi \rightarrow \psi|_S = (U - |\varphi|_S) \cup |\psi|_S,$
$|\varphi \equiv \psi|_S = |\varphi|_S \rightarrow |\psi|_S \cap |\varphi|_S \rightarrow |\psi|_S.$

Thus, the meaning of the formula φ is the set of all objects having the property expressed by the formula φ, or the meaning of the formula φ is the description in the *KR*-language of the set objects $|\varphi|$.

A formula φ is said to be *true* in a KR-system S, denoted $\models_S \varphi$, iff $|\varphi|_S = U$, i.e., the formula is satisfied by all objects of the universe in the system S. Formulas φ and ψ are equivalent in S iff $|\varphi|_S = |\psi|_S$.

Proposition 3.3 *The following are the simple properties of the meaning of a formula.*
$\models_S \varphi$ *iff* $|\varphi|_S = U,$
$\models_S \sim\varphi$ *iff* $|\varphi|_S = \emptyset,$
$\varphi \rightarrow \psi$ *iff* $|\psi|_S \subseteq |\psi|_S,$
$\varphi \equiv \psi$ *iff* $|\psi|_S = |\psi|_S.$

To deal with deduction in *DL*, we need suitable axioms and inference rules. Here, the axioms will correspond closely to the axioms of classical propositional logic, but some specific axioms for the specific properties of knowledge representation systems are also needed. The only inference rule will be *modus ponens*.

We will use the following abbreviations:

$$\varphi \wedge \sim \varphi =_{\text{def}} 0 \text{ and } \varphi \vee \sim \varphi =_{\text{def}} 1.$$

A formula of the form $(a_1, v_1) \wedge (a_2, v_2) \wedge \cdots \wedge (a_n, v_n)$, where $v_{ai} \in V_a$, $P = \{a_1, a_2, \ldots, a_n\}$, and $P \subseteq A$, is called a *P-basic formula* or in short *P*-formula. An atomic formula is called an *A-basic formula* or in short a basic formula.

Let $P \subseteq A$, φ be a P-formula, and $x \in U$. The set of all A-basic formulas satisfiable in the knowledge representation system $S = (U, A)$ is called the *basic knowledge* in S. We write $\underset{S}{\Sigma}(P)$ to denote the disjunction of all P-formulas satisfied in S. If $P = A$, then $\underset{S}{\Sigma}(A)$ is called the *characteristic formula* of S.

The knowledge representation system can be represented by a data table. Its columns are labeled by attributes, and its rows are labeled by objects. Thus, each row in the table is represented by a certain A-basic formula, and the whole table is represented by the set of all such formulas. In *DL*, instead of tables, we can use sentences to represent knowledge. There are specific axioms of *DL*:

1. $(a, v) \wedge (a, u) \equiv 0$ for any $a \in A, u, v \in V$, and $v \neq u$.
2. $\bigvee_{v \in V_a} \equiv 1$ for every $a \in A$.
3. $\sim (a, v) \equiv \bigvee_{a \in V_a, u \neq v} (a, u)$ for every $a \in A$.

We say that a formula φ is derivable from a set of formulas Ω, denoted $\Omega\varphi$, iff it is derivable from the axioms and formulas of Ω by a finite application of modus ponens. Formula φ is a theorem of *DL*, denoted φ, if it is derivable from the axioms only. A set of formulas Ω is consistent iff the formula $\varphi \wedge \sim \varphi$ is not derivable from Ω.

Note that the set of theorems of *DL* is identical with the set of theorems of classical propositional logic with specific axioms (1)–(3), in which negation can be eliminated.

Formulas in the KR-language can be represented in a special form called a *normal form*, which is similar to that in classical propositional logic. Let $P \subseteq A$ be a subset of attributes and let φ be a formula in the KR-language.

We say that φ is in a P-normal form in S, in short in P-normal form, iff either φ is 0 or φ is 1, or φ is a disjunction of non-empty P-basic formulas in S. (The formula φ is non-empty if $|\varphi| \neq \emptyset$).

A-normal form will be referred to as normal form. The following is an important property in the *DL*-language.

Proposition 3.4 *Let φ be a formula in a DL-language, and let P contain all attributes occurring in φ. Moreover, assume axioms (1)–(3) and the formula $\sum_S(A)$. Then, there is a formula ψ in the P-normal form such that $\varphi \equiv \psi$.*

Definition 3.6 A translation τ from the propositional constant L to an interpretation of a rough set language $\mathscr{L}_{\mathrm{RS}}$ of atomic expressions in the *KR*-system *S* is combined with $\neg, \vee, \wedge$ and $\rightarrow$ such that

$\tau(|\varphi|_S) = |(a, v)|_S$,
$\tau(| \sim \varphi|_S) = -\tau(|\varphi|_S)$,
$\tau(|\varphi \vee \psi|_S) = \tau(|\varphi|_S) \cup \tau(|\psi|_S)$,
$\tau(|\varphi \wedge \psi|_S) = \tau(|\varphi|_S) \cap \tau(|\psi|_S)$,
$\tau(|\varphi \rightarrow \psi|_S) = -\tau(|\varphi|_S) \cup \tau(|\psi|_S)$,
$\tau(|\varphi \equiv \psi|_S) = (\tau(|\varphi|_S) \cap \tau(|\psi|_S)) = (-\tau(|\varphi|_S) \cap -\tau(|\psi|_S))$.

Let φ be an atomic formula of the DL-language, $R \in C \cup D$ an equivalence relation, X any subset of U, and a valuation v of propositional variables. Then, the truth-values of φ are defined as follows:

$$\|\varphi\|^v = \begin{cases} \mathbf{t} \ \textit{if} \ |\varphi|_S \subseteq POS_R(U/X) \\ \mathbf{f} \ \textit{if} \ |\varphi|_S \subseteq NEG_R(U/X) \end{cases}$$

This shows that decision logic is based on bivalent logic. In the next section, an interpretation of decision logic based on three-valued logics will be discussed.

3.3 Belnap's Four-Valued Logic

Belnap [5, 6] first claimed that an inference mechanism for a database should employ a certain four-valued logic. The important point in Belnap's system is that we should deal with both incomplete and inconsistent information in databases.

To represent such information, we need a four-valued logic since classical logic is not appropriate for the task. Belnap's four-valued semantics can in fact be viewed as an intuitive description of the internal states of a computer.

In Belnap's four-valued logic **B4**, four kinds of truth-values are used from the set **4** = {**T**, **F**, **N**, **B**}. These truth-values can be interpreted in the context of a computer, namely **T** means just told True, **F** means just told False, **N** means told neither True nor False, and **B** means told both True and False. Intuitively, **N** can be equated as $\emptyset$, and **B** as overdefined.

Belnap outlined a semantics for **B4** using logical connectives. Belnap's semantics uses a notion of *set-ups* mapping atomic formulas into **4**. A *set-up* can then be extended for any formula in **B4** in the following way:

$$s(A \,\&\, B) = s(A) \,\&\, s(B),$$
$$s(A \vee B) = s(A) \vee s(B),$$
$$s(\sim A) = \sim s(A).$$

Belnap also defined a concept of entailments in **B4**. We say that A entails B just in case for each assignment of one of the four values to variables, the value of A does not exceed the value of B in B4, i.e., $s(A) \leq s(B)$ for each set-up s.

Here, $\leq$ is defined as $\mathbf{F} \leq \mathbf{B}$, $\mathbf{F} \leq \mathbf{N}$, $\mathbf{B} \leq \mathbf{T}$, $\mathbf{N} \leq \mathbf{T}$. Belnap's four-valued logic in fact coincides with the system of *tautological entailments* due to Anderson and Belnap [3]. Belnap's logic **B4** is one of the paraconsistent logics capable of tolerating contradictions. Belnap also studied the implications and quantifiers in **B4** in connection with question-answering systems. However, we will not go into detail here.

The structure that consists of these four elements and the five basic operators is usually called **B4**.

Designated elements and models: The next step in using **B4** for reasoning is to choose its set of *designated* elements. The obvious choice is $\mathcal{D} = \{\mathbf{T}, \mathbf{B}\}$ since both values intuitively represent a formula known to be true.

The set $\mathcal{D}$ has the property that $a \wedge b \in \mathcal{D}$ iff both a and b are in $\mathcal{D}$, while $a \vee b \in \mathcal{D}$ iff either a or b is in $\mathcal{D}$. From this point, various semantics notions are defined on **B4** as natural generalizations of similar classical notions.

3.4 Rough Sets and Partial Semantics

Partial semantics for classical logic has been studied by van Benthem [25] in the context of the *semantic tableaux*. This insight can be generalized to study consequence relations in terms of a Gentzen-type sequent calculus. To handle an aspect of vagueness on the decision logic, the forcing relation for the partial interpretation is defined as a four-valued semantic.

As the proposed approach can replace the base bivalent logic of decision logic with many-valued logics, alternative versions of decision logic based on many-valued logics are obtained.

The model $\mathscr{S}$ of decision logic based on four-valued semantics consists of a universe U for the language L and an assignment function s that provides an interpretation for L.

For the domain $|\mathscr{S}|$ of the model $\mathscr{S}$, a subset is defined by $S = \langle S^+, S^- \rangle$. The first term of the ordered pair denotes the set of n-tuples of elements of the universe that *verify* the relation S, whereas the second term denotes the set of n-tuples that *falsify* the relation.

The interpretation of the propositional variables of L for the model $\mathscr{S}$ is given by $S_{\mathscr{S}} = \langle (S)^+_{\mathscr{S}}, (S)^-_{\mathscr{S}} \rangle$. An *interpretation function* for a domain $|\mathscr{S}|$ in the standard way as a function s with domain L such that $s(x) \in |\mathscr{S}|^n$ if S is a relation symbol.

We need two interpretation functions for each model here; a model for partial logic for a predicate symbol is a triple $\langle |\mathscr{S}|, s^+, s^- \rangle$, where s^+ and s^- are interpretation functions for $|\mathscr{S}|$.

The denotation of a relation symbol consists of those tuples for which it is *true* that they stand in the relation; the antidenotation consists of the tuples for which this is *false*.

As before, truth and falsity are neither true nor false, or it may be both true and false that some tuple stands in a certain relation. The following definition is modified from [16].

Definition 3.7 (Partial Relation) An n-ary *partial relation* S on the domain $|\mathscr{S}_1|, \ldots, |\mathscr{S}_n|$ is a tuple $\langle S^+, S^- \rangle$ of the relations $S^+, S^- \subseteq |\mathscr{S}_1| \times \cdots \times |\mathscr{S}_n|$. The relation S^+ is called S's *denotation*; S^- is called S's *antidenotation*, $|\mathscr{S}_1| \times \cdots \times |\mathscr{S}_n| / (S^+ \cup S^-)$ its gap, and $S^+ \cap S^-$ its glut.

A partial relation is *coherent* if its glut is empty, *total* if its gap is empty, *incoherent* if it is not coherent and *classical* if it is both coherent and total. A unary partial relation is called a *partial set*.

Definition 3.8 (Partial Operation for 4) Let $S_1 = \langle S_1^+, S_1^- \rangle$ and $S_2 = \langle S_2^+, S_2^- \rangle$ be partial relations. Define

$$
\begin{aligned}
-S_1 &:= \langle S_1^+, S_1^- \rangle \ (\textit{partial complementation}),\\
S_1 \cap S_2 &:= \langle S_1^+ \cap S_2^-, S_1^+ \cup S_2^- \rangle \ (\textit{partial intersection}),\\
S_1 \cup S_2 &:= \langle S_1^+ \cup S_2^-, S_1^+ \cap S_2^- \rangle \ (\textit{partial union}),\\
S_1 \subseteq S_2 &:= \langle S_1^+ \subseteq S_2^-, S_1^+ \subseteq S_2^- \rangle \ (\textit{partial inclusion}).
\end{aligned}
$$

Partial inclusion means S_1 approximates S_2.

Let A be some set of partial relations; then, the following properties hold:

$$\bigcap A := \langle \bigcap \{S^+ \mid S \in A\}, \bigcup \{S^- \mid S \in A\} \rangle,$$
$$\bigcup A := \langle \bigcup \{S^+ \mid S \in A\}, \bigcap \{S^- \mid S \in A\} \rangle.$$

To handle three-valued and four-valued logic in a unified manner, we adopt the four-value interpretation by Belnap [5].

Let $\mathbf{4} = \{\mathbf{T}, \mathbf{F}, \mathbf{N}, \mathbf{B}\}$ be the truth-values for the four-valued semantics of L, where each value is interpreted as true, false, neither true nor false, and both true and false.

A model $\mathscr{S}$ determines a four-valued assignment v on atomic formula in the following way:

$$\|\varphi\|^v = \begin{Bmatrix} \mathbf{T} \\ \mathbf{F} \\ \mathbf{N} \\ \mathbf{B} \end{Bmatrix} \textit{if } |\varphi, \sim\varphi|_S \cap S = \begin{Bmatrix} \{\varphi\} \\ \{\sim\varphi\} \\ \{\emptyset\} \\ \{\varphi, \sim\varphi\} \end{Bmatrix}.$$

Then, the truth-values of φ on $S = (U, A)$ is defined as follows:

$$\|\varphi\|^v = \begin{cases} \mathbf{T} \textit{ if } |\varphi|_S \subseteq POS_R(U/X) \\ \mathbf{F} \textit{ if } |\varphi|_S \subseteq NEG_R(U/X) \\ \mathbf{N} \textit{ if } |\varphi|_S \nsubseteq POS_R(U/X) \cup NEG_R(U/X) \\ \mathbf{B} \textit{ if } |\varphi|_S \subseteq BN_R(U/X) \end{cases}$$

Definition 3.9 (Partial Model) A partial model for a propositional DL-language L is a tuple $\mathscr{M} = (\mathscr{T}, \mathscr{D}, \mathscr{O})$, where

- $\mathscr{T}$ is a non-empty set of truth-values.
- $\emptyset \subset \mathscr{D} \subseteq \mathscr{T}$ is the set of designated values.
- For every *n-ary connective* $\diamond$ of L, $\mathscr{O}$ includes a corresponding n-ary function $\tilde{\diamond}$ from $\mathscr{T}^n$ to $\mathbf{4}$.

Let W be the set of *well-formed formulas* of L. A (legal) valuation in a Partial Model $\mathscr{S}$ is a function $V : W \to \mathbf{4}$ that satisfies the following condition:

$$V(\diamond(\psi_1, \ldots, \psi_n)) \in \tilde{\diamond}(V(\psi_1), \cdots, V(\psi_n))$$

for every n-ary connective $\diamond$ of L and any $\psi_1, \ldots, \psi_n \in W$.

Let $\mathscr{V}_M$ denote the set of all valuations in the partial model $\mathscr{D}$. The notions of satisfaction under a valuation, validity, and consequence relation are defined as follows:

- A formula $\varphi \in W$ is satisfied by a valuation $v \in \mathscr{V}_M$, in symbols, $\mathscr{M} \models_v \varphi$, $v(\varphi) \in \mathscr{D}$.
- A sequent $\Sigma = \Gamma \Rightarrow \Delta$ is satisfied by a valuation $v \in \mathscr{V}_M$, in symbols, $\mathscr{M} \models_v \Sigma$, iff either v does not satisfy some formula in Γ or v satisfies some formula in Δ.
- A sequent Σ is valid, in symbols, $\models \Sigma$, if it is satisfied by all valuations $V \in \mathscr{V}_M$.

- The consequence relation on W defined by $\mathscr{M}$ is the relation $\mathscr{M} \vdash$ on sets of formulas in W such that, for any $T, S \subseteq W$, $T \vdash_{\mathscr{M}} S$ iff there exist finite sets $\Gamma \subseteq T$, $\Delta \subseteq S$ such that the sequent $\Gamma \Rightarrow \Delta$ is valid.

Definition 3.10 (Tarski truth definition for partial propositional logic) Let L be a set of propositional constants and let $v : P \to \{\mathbf{T}, \mathbf{F}, \mathbf{N}, \mathbf{B}\}$ be a (*valuation*) *function.*

$$\|p\|^v = v(p) \text{ if } p \in P.$$

The truth-values of φ on the information system $S = (U, A)$ are represented by forcing relations as follows:

$$\|\varphi\|^v = \mathbf{T} \textit{ iff } \mathscr{M} \models_v^+ \varphi \textit{ and } \mathscr{M} \not\models_v^- \varphi,$$
$$\|\varphi\|^v = \mathbf{F} \textit{ iff } \mathscr{M} \not\models_v^+ \varphi \textit{ and } \mathscr{M} \models_v^- \varphi,$$
$$\|\varphi\|^v = \mathbf{N} \textit{ iff } \mathscr{M} \not\models_v^+ \varphi \textit{ and } \mathscr{M} \not\models_v^- \varphi,$$
$$\|\varphi\|^v = \mathbf{B} \textit{ iff } \mathscr{M} \models_v^+ \varphi \textit{ and } \mathscr{M} \models_v^- \varphi.$$

A semantic relation for the model $\mathscr{M}$ is defined following [8, 16, 25]. The truth and falsehood of a formula of the DL-language are defined in a model $\mathscr{M}$.

The truth (denoted by $\models_v^+$) and the falsehood (denoted by $\models_v^-$) of the formulas of the decision logic in $\mathscr{M}$ are defined inductively.

Definition 3.11 The semantic relations of $\mathscr{M} \models_v^+ \varphi$ and $\mathscr{M} \models_v^- \varphi$ are defined as follows:

$\mathscr{M} \models_v^+ \varphi$ *iff* $\varphi \in M^+$,
$\mathscr{M} \models_v^- \varphi$ *iff* $\varphi \in M^-$,
$\mathscr{M} \models_v^+ \sim \varphi$ *iff* $\mathscr{M} \models_v^- \varphi$,
$\mathscr{M} \models_v^- \sim \varphi$ *iff* $\mathscr{M} \models_v^+ \varphi$,
$\mathscr{M} \models_v^+ \varphi \vee \psi$ *iff* $\mathscr{M} \models_v^+ \varphi$ *or* $\mathscr{M} \models_v^+ \psi$,
$\mathscr{M} \models_v^- \varphi \vee \psi$ *iff* $\mathscr{M} \models_v^- \varphi$ *and* $\mathscr{M} \models_v^- \psi$,
$\mathscr{M} \models_v^+ \varphi \wedge \psi$ *iff* $\mathscr{M} \models_v^+ \varphi$ *and* $\mathscr{M} \models_v^+ \psi$,
$\mathscr{M} \models_v^- \varphi \wedge \psi$ *iff* $\mathscr{M} \models_v^- \varphi$ *or* $\mathscr{M} \models_v^- \psi$,
$\mathscr{M} \models_v^+ \varphi \to \psi$ *iff* $\mathscr{M} \models_v^- \varphi$ *or* $\mathscr{M} \models_v^+ \psi$,
$\mathscr{M} \models_v^- \varphi \to \psi$ *iff* $\mathscr{M} \models_v^+ \varphi$ *and* $\mathscr{M} \models_v^- \psi$.

The symbol $\sim$ denotes strong negation, in which $\sim$ is interpreted as true if the proposition is false.

Since validity in **B4** is defined in terms of truth preservation, the set of designated values is $\{\mathbf{T}, \mathbf{B}\}$ of **4**. We assume that an interpretation of **B4** satisfies the following constraint.

Definition 3.12 (Exclusion and Exhaustion)
Exclusion: model $\mathscr{M}$ is *exclusion iff* $S^+ \cap S^- = \emptyset$.
Exhaustion: model $\mathscr{M}$ is *exhaustion iff* $S^+ \cup S^- = S$.

The model $\mathscr{M}$ is *consistent* if and only if $S^+ \cap S^- = \emptyset$. The relational domains of general models are closed under the operations $\cap$, $\cup$.

The natural operation on the set of truth combinations $\mathbf{4} = \{\mathbf{T}, \mathbf{F}, \mathbf{N}, \mathbf{B}\}$ that we have defined in the previous section can be extended to the class of partial relations.

Definition 3.13 A model of **B4** for L is a pair $M = (S, |\cdot|)$, where S is a non-empty set, and $|\cdot|$ is an interpretation of a propositional symbol, with $|p| : S_n \rightarrow \mathbf{4}$ for any $p \in P_n, n \leq 0$.

Example 3.1 Suppose the decision table below where the condition and decision attributes are not considered.

$U = \{x_1, x_2, x_3, x_4, x_5, x_6, x_7, x_8\}$
Attribute: $C = \{c1, c2, c3, c4\}$
$c_1 = \{x_1, x_4, x_8\}, c_2 = \{x_2, x_5, x_7\}, c_3 = \{x_3\},$
$c_4 = \{x_6\}$
$U/C = c_1 \cup c_2 \cup c_3 \cup c_4$
Any subset $X = \{x_3, x_6, x_8\}$
$POS_C(X) = c_3 \cup c_4 = \{x_3, x_6\}$
$BN_C(X) = c_1 = \{x_1, x_4, x_8\}$
$NEG_C(X) = c_2 = \{x_2, x_5, x_7\}$
The evaluation of the truth-values of the formulas is as follows:
$If\, |C_{c3}|_S \subseteq POS_C(X)\ then\ \|C_{c3}\|^v = \mathbf{T},$
$If\, |C_{c2}|_S \subseteq NEG_C(X)\ then\ \|C_{c2}\|^v = \mathbf{F},$
$If\, |C_{c2}|_S \not\subseteq POS_C(X) \cup NEG_C(X)\ then\ \|C_{c2}\|^v = \mathbf{N},$
$If\, |C_{c1}|_S \subseteq BN_C(X)\ then\ \|C_{c1}\|^v = \mathbf{B}.$

Example 3.2 Consider Table 3.1 again. Assume that a, b, and c are condition attributes and d and e are decision attributes. Decision rules 1 and 5 are inconsistent. This means that 1 and 5 can be considered to have non-deterministic values, e.g., **N** or **B** respectively.

3.5 Consequence Relation and Sequent Calculus

Partial semantics for classical logic is closely related to the interpretation of the Beth tableau [23]. Van Benthem [25] suggested the relationship of the consequence relation to a Gentzen sequent calculus. We replace the bivalent logic of the decision logic with many-valued logics based on partial semantics.

We begin by recalling the basic idea of the *Beth tableau*. The Beth tableaux proves $X \rightarrow Y$ by constructing a counterexample of $X\ \&\ \sim Y$. The Beth tableaux has several partial features.

For instance, there may be counterexamples even if a branch remains open. This insight led van Benthem [25] to work out partial semantics for classical logic.

Here, we describe sequent calculi shortly. For sequent calculi, formulas are constructed from the propositional variables and logical connectives, e.g., $\sim$, $\neg$, $\wedge$, $\vee$, and $\rightarrow$. Capital letters $A, B, \ldots$ are used for formulas, and Greek capital letters Γ, Δ are used for finite sequences of formulas. A sequent is an expression of the form $\Gamma \Rightarrow A$.

We introduce some concepts of sequent calculi. If a sequent $\Gamma \Rightarrow A$ is provable in a system S, then we write $S \vdash \Gamma \Rightarrow A$. A rule R of inference holds for a system S if the following condition is satisfied.

For any instance of the following sequent of R, if $S \vdash \Gamma_i \Rightarrow A_i$ for all i, then $S \vdash \Delta \Rightarrow B$.

$$\frac{\Gamma_1 \Rightarrow A_1 \quad \cdots \quad \Gamma_n \Rightarrow A_n}{\Delta \Rightarrow B}$$

Moreover, R is said to be derivable in S if there is a derivation from $\Gamma_1 \Rightarrow A_1, \ldots, \Gamma_n \Rightarrow A_n$ to $\Delta \Rightarrow B$ in S.

To accommodate the Gentzen system to partial logics, we need some concepts of partial semantics. In the Beth tableau, It is assumed that V is a partial valuation function assigning the values 0 or 1 to an atomic formula p. We can then set $V(p) = 1$ for p on the left-hand side and $V(p) = 0$ for p on the right-hand side in an open branch of the tableau.

To deal with an uncertain concept in many-valued semantics, we need to introduce the consequence relation [24]. *Pre* and *Cons* represent the sequent premise and conclusion, respectively, and 1 represents true and 0 false. First, we define the following concept of consequence relation C1.

$$\text{(C1) for all } V \text{, if } V(Pre) = 1 \text{, then } V(Cons) = 1.$$

In (C1), if *Pre* is evaluated as 1, then *Cons* preserves 1. Here, we define a classical Gentzen system.

Definition 3.14 (Sequent Calculus for Classical Propositional Logic CL) The sequent calculus for the classical propositional logic **CL** is defined as follows:
Axiom: (ID) $A \Rightarrow A$
Sequent rules:

$$(Weakening)\ \frac{\Gamma \Rightarrow \Delta}{\Gamma' \Rightarrow \Delta'}\ (\Gamma \subset \Gamma', \Delta \subset \Delta') \qquad (Cut)\ \frac{\Gamma, A \Rightarrow \Delta \quad \Gamma \Rightarrow A, \Delta}{\Gamma \Rightarrow \Delta}$$

$$(\sim R)\frac{A, \Gamma \Rightarrow \Delta}{\Gamma \Rightarrow \Delta, \sim A} \qquad (\sim L)\frac{\Gamma \Rightarrow \Delta, A}{\sim A, \Gamma \Rightarrow \Delta}$$

$$(\wedge R)\frac{\Gamma \Rightarrow \Delta, A \quad \Gamma \Rightarrow \Delta, B}{\Gamma \Rightarrow \Delta, A \wedge B} \qquad (\wedge L)\frac{A, B, \Gamma \Rightarrow \Delta}{A \wedge B, \Gamma \Rightarrow \Delta}$$

$$(\vee R)\frac{\Gamma \Rightarrow \Delta, A, B}{\Gamma \Rightarrow \Delta, A \vee B} \qquad (\vee L)\frac{A, \Gamma \Rightarrow \Delta \quad B, \Gamma \Rightarrow \Delta}{A \vee B, \Gamma \Rightarrow \Delta}$$

$$(\to R)\frac{\Gamma \Rightarrow \Delta, \sim A, B}{\Gamma \Rightarrow \Delta, A \to B} \quad (\to L)\frac{\sim A, \Gamma \Rightarrow \Delta \quad B, \Gamma \Rightarrow \Delta}{A \to B, \Gamma \Rightarrow \Delta}$$

Theorem 3.1 *The logic for C1 is axiomatized by the Gentzen sequent calculus CL.*

Proof See [2, 23, 25].

Next, we define the sequent calculus GC1 for C1 that can be obtained by adding the following rules to CL without ($\sim R$) such as CL\{($\sim R$)}, where, "\" implies that the rule following "\" is excluded:

$$(\sim\sim R)\frac{\Gamma \Rightarrow \Delta, A}{\Gamma \Rightarrow \Delta, \sim\sim A} \quad (\sim\sim L)\frac{A, \Gamma \Rightarrow \Delta}{\sim\sim A, \Gamma \Rightarrow \Delta}$$

$$(\sim \wedge R)\frac{\Gamma \Rightarrow \Delta, \sim A, \sim B}{\Gamma \Rightarrow \Delta, \sim (A \wedge B)} \quad (\sim \wedge L)\frac{\sim A, \Gamma \Rightarrow \Delta \quad \sim B, \Gamma \Rightarrow \Delta}{\sim (A \wedge B), \Gamma \Rightarrow \Delta}$$

$$(\sim \vee R)\frac{\Gamma \Rightarrow \Delta, \sim A, \quad \Gamma \Rightarrow \Delta, \sim B}{\Gamma \Rightarrow \Delta, \sim (A \vee B)} \quad (\sim \vee L)\frac{\sim A, \sim B, \Gamma \Rightarrow \Delta}{\sim (A \vee B), \Gamma \Rightarrow \Delta}$$

It is worth noting that the three-valued logic by Kleene has no tautology. Thus, to define a consequence relation, a tableaux system for a three-valued logic is formalized [2, 25]. Then, the consequence relation C2 is defined as follows:

$$\text{(C2) for all } V, \text{if } V(Pre) = 1, \text{then } V(Cons) \neq 0.$$

C2 is interpreted as *exclusion*; then, the consequence relation C2 is regarded for Kleene's strong three-valued logic K_3.

As the semantics for (C2), we define the extension of the valuation function $V^{C2}(p)$ with $v(p)$ for an atomic formula p as follows:

$\mathbf{T} =_{\text{def}} V^{C2}(p) = \{1\} =_{\text{def}} v(p) = 1 \; and \; v(p) \neq 0,$
$\mathbf{F} =_{\text{def}} V^{C2}(p) = \{0\} =_{\text{def}} v(p) = 0 \; and \; v(p) \neq 1,$
$\mathbf{N} =_{\text{def}} V^{C2}(p) = \{\} =_{\text{def}} v(p) \neq 1 \; and \; v(p) \neq 0.$

The interpretation of C2 by the partial semantics is given as follows:

Definition 3.15 $\Gamma \models_s \varphi$ iff there is no φ that is not **F** under V^{C2} (in the three-valued {**T**, **F**, **N**}) and for all $\gamma \in \Gamma$, γ is **T** under V^{C2}.

The Gentzen-type sequent calculus GC2 axiomatizes C2 [2, 25]. We are now in a position to define GC2. For GC2, the principle of explosion (ex falso quodlibet (EFQ)), defined below, is added to TG1\{($\sim L$)}.

(EFQ) $A, \sim A \Rightarrow$

Definition 3.16 The sequent calculus GC2 is defined as follows:
GC2 := {(ID), ($Weakening$), (Cut), (EFQ), ($\wedge R$), ($\wedge L$), ($\vee R$), ($\vee L$), ($\to R$), ($\to L$), ($\sim\sim R$), ($\sim\sim L$), ($\sim \wedge R$), ($\sim \wedge L$), ($\sim \vee R$), ($\sim \vee L$)}.

GC2 can be interpreted as *truth preserving* with the matrix of a three-valued logic defined as $\langle\{T, F, N\}, \{T\}, \{\sim, \vee, \wedge, \rightarrow\}\rangle$. For the rule ($\sim L$) obtained from (EFQ), GC2 and GC1 are equivalent.

Theorem 3.2 *GC2 = GC1.*

Proof (EFQ) can be considered as ($\sim L$); then, double negation and the de Morgan laws in GC2 are obtained.

In the classical interpretation of CL, the law of excluded middle (LEM) holds but not in C2. Then, the consequence relation C2 is axiomatized as GC2.

Theorem 3.3 *C2 can be axiomatized by the sequent calculus GC2.*

Proof See [2, 25].

Theorem 3.4 *In the model for C2, $\mathscr{S}$, DL-language L, and formula φ, it is not the case that $\mathscr{M} \models_v^+ \varphi$ and $\mathscr{M} \models_v^- \varphi$ hold.*

Proof Only the proof for $\sim$ and $\wedge$ will be provided. It can be carried out by induction on the complexity of the formula. The condition of *consistent* implies that it is not the case that $\varphi \in \mathscr{S}^+$ and $\varphi \in \mathscr{S}^-$. Then, it is not the case that $\mathscr{M} \models_v^+ \varphi$ and $\mathscr{M} \models_v^- \varphi$.

$\sim$: We assume that $\mathscr{M} \models_v^+ \sim \varphi$ and $\mathscr{M} \models_v^- \sim \varphi$ hold. Then, it follows that $\mathscr{M} \models_v^+ \varphi$ and $\mathscr{M} \models_v^- \varphi$. This is a contradiction.

$\wedge$: We assume that $\mathscr{M} \models_v^- \varphi \wedge \psi$ and $\mathscr{M} \models_v^+ \varphi \wedge \psi$ hold. Then, it follows that $\mathscr{M} \models_v^+ \varphi$, $\mathscr{M} \models_v^+ \psi$ and either $\mathscr{M} \models_v^- \varphi$ or $\mathscr{M} \models_v^- \psi$. In either case, there is a contradiction.

Next, we provide another consequence relation with a different interpretation for the third-value below.

$$(C3)\text{for all } V, \text{if } V(Pre) \neq 0, \text{then } V(Cons) = 0.$$

C3 is interpreted as *exhaustion*, then the consequence relation C3 is for Logic for Paradox [21]. As the semantics for C3, we define the extension of the valuation function $V^{C3}(p)$ with $v(p)$ for an atomic formula p as follows:

$$\begin{aligned}
&\mathbf{T} =_{\text{def}} V^{C3}(p) = \{1\} =_{\text{def}} v(p) = 1 \; and \; v(p) \neq 0,\\
&\mathbf{F} =_{\text{def}} V^{C3}(p) = \{0\} =_{\text{def}} v(p) = 0 \; and \; v(p) \neq 1,\\
&\mathbf{B} =_{\text{def}} V^{C3}(p) = \{1, 0\} =_{\text{def}} v(p) = 1 \; and \; v(p) = 0.
\end{aligned}$$

The interpretation of C3 by the partial semantics is given as follows:

Definition 3.17 $\Gamma \models_v \varphi$ iff there is φ that is $\mathbf{T}$ under V^{C3} (in the three-valued $\{\mathbf{T}, \mathbf{F}, \mathbf{B}\}$) and for all $\gamma \in \Gamma$, γ is not $\mathbf{F}$ under V^{C3}.

The Gentzen sequent calculus GC3 is obtained from GC2, replacing EFQ with LEM (*the law of excluded middle*) as an axiom:

$$(LEM) \Rightarrow A, \sim A$$

Definition 3.18 The sequent calculus GC3 is defined as follows:
GC3 := $\{(ID), (Weakening), (Cut), (LEM), (\wedge R), (\wedge L), (\vee R), (\vee L), (\rightarrow R), (\rightarrow L), (\sim\sim R), (\sim\sim L), (\sim \wedge R), (\sim \wedge L), (\sim \vee R), (\sim \vee L)\}$.

Theorem 3.5 *C3 can be axiomatized by the Gentzen calculus GC3.*

Proof GC3 can be obtained by deriving double negation and two de Morgan laws in GC3. The $(\sim R)$ rule can be provided as LEM.

Next, we extend consequence relation C4 as follows:

$$(C4)\text{for all } V, \text{if } V(Pre) \neq 0, \text{then } V(Cons) \neq 0.$$

C4 is regarded as a four-valued logic since it allows for an inconsistent valuation. We are now in a position to define Belnap's four-valued logic **B4**.

As the semantics for GC4, Belnap's **B4** is adopted here. We define the extension of the valuation function $V^{C4}(p)$ with $v(p)$ for an atomic formula p as follows:

T $=_{\text{def}} V^{C4}(p) = \{1\} =_{\text{def}} v(p) = 1$ *and* $v(p) \neq 0$,
F $=_{\text{def}} V^{C4}(p) = \{0\} =_{\text{def}} v(p) = 0$ *and* $v(p) \neq 1$,
N $=_{\text{def}} V^{C4}(p) = \{\} =_{\text{def}} v(p) \neq 1$ *and* $v(p) \neq 0$,
B $=_{\text{def}} V^{C4}(p) = \{1, 0\} =_{\text{def}} v(p) = 1$ *and* $v(p) = 0$.

The interpretation of C4 by the partial semantics is given as follows:

Definition 3.19 $\Gamma \models_v \varphi$ *iff* there is no φ that is not **F** under V^{C4} (in **4**) and for all $\gamma \in \Gamma$, γ is not **F** under V^{C4}.

Definition 3.20 The sequent calculus GC4 is defined as follows:
GC4 := $\{(ID), (Weakening), (Cut), (\wedge R), (\wedge L), (\vee R), (\vee L), (\sim\sim R), (\sim\sim L), (\sim \wedge R), (\sim \wedge L), (\sim \vee R), (\sim \vee L)\}$.

Theorem 3.6 *C4 can be axiomatized by the sequent calculus GC4.*

Proof GC4 can be obtained by deriving double negation and two de Morgan laws in GC4. The $(F \sim)$ rule can be provided as LEM.

3.6 Extension of Many-Valued Semantics

We introduce some three-valued logics and provide relationships and properties between the consequence relations.

Kleene's strong three-valued logic: Kleene proposed three-valued logics to deal with *undecidable* sentences in connection with recursive function theory [12]. Thus, the third truth-value can be interpreted as undecided in the strong Kleene logic K_3, which is of special interest to describe a machine's computational state.

Table 3.2 Truth tables for K3

$\sim$	
T	F
F	T
N	N

$\wedge$	T	F	N
T	T	F	N
F	F	F	F
N	N	F	N

$\vee$	T	F	N
T	T	T	T
F	T	F	N
N	T	N	N

$\rightarrow$	T	F	N
T	T	F	N
F	T	T	T
N	T	N	N

K3 can give a truth value to a compound sentence even if some of its parts have no truth value. Kleene also proposed the weak three-valued logic in which the whole sentence is undecided if any component of a compound sentence is undecided.

The truth tables for K3 are defined as Table 3.2:

The implication $\rightarrow$ can be defined in the following way:

$$A \rightarrow B =_{\text{def}} \sim A \vee B$$

The axiomatization of K3 by a Gentzen-type sequent calculus can be found in the literature [24].

Let $\models$ be the consequence relation of K3. Then, we have the following Gentzen-type sequent calculus GK3 for K3, which contains an axiom of the form:

$$X \models Y \text{ if } X \cap Y \neq \emptyset$$

and the rules (Weakening), (Cut), and

$$A \models \sim\sim A, \qquad \sim\sim A \models A, \qquad A, \sim A \models,$$
$$A, B \models A \wedge B, \qquad A \wedge B \models A, \qquad A \wedge B \models B,$$
$$\sim A \models \sim (A \wedge B), \qquad \sim B \models \sim (A \wedge B),$$
$$\sim (A \wedge B) \models \sim A, \sim B.$$

GC2 is considered as Kleene's strong three-valued logic K3. The implication of K3 does not satisfy the deduction theorem. In addition, $A \rightarrow A$ is not a theorem in K3.

Theorem 3.7 $\models_{K3} = \vdash_{C2}$, *where* $\models_{K3}$ *denotes the consequence relation of* K_3.

Proof By induction on K3 and GC2. It is easy to transform each proof of K3 into GC2. The converse transformation can also be presented.

Łukasiewicz three-valued logic: Łukasiewicz's three-valued logic was proposed in 1920 in order to interpret a future contingent statement in which the third truth-value can be read as indeterminate or possible [14]. Thus, in Łukasiewicz's three-valued logic L_3, neither the law of excluded middle nor the law of non-contradiction holds.

Table 3.3 Truth table for L_3

$\supset$	T	F	N
T	T	F	N
F	T	T	T
N	T	N	T

The difference between K_3 and L_3 lies in the interpretation of implication, as the truth table (Table 3.3) indicates.

It is also possible to describe the Hilbert presentation of L_3. Let $\supset$ be the Łukasiewicz implication. Then, we can show the following axiomatization of L_3 due to Wajsberg. It has been axiomatized by Wajsberg (cf. [24]) using a language based on $(\vee, \supset, \sim)$, the modus ponens rule and the following axioms:

(W1) $(p \supset q) \supset ((p \supset r) \supset (p \supset r))$,
(W2) $(\sim p \supset \sim q) \supset (q \supset p))$,
(W3) $(((p \supset \sim p) \supset p) \supset p)$.

They are closed under the rules of substitution and *modus ponens*. Unlike in K_3, $A \supset A$ is a theorem in L_3. It is noted, however, that the philosophical motivation of L_3 in connection with Aristotelian logic can be challenged.

The definition of the semantic relation for the implication of L_3 is obtained by replacing the implication in Definition 3.11 with the following definition:

$\mathcal{M} \models_v^+ \varphi \rightarrow \psi$ *iff* $\mathcal{M} \models_v^- \varphi$ *or* $\mathcal{M} \models_v^+ \psi$ or $(\mathcal{M} \not\models_v^+ \varphi$ *and* $\mathcal{M} \not\models_v^- \varphi$ *and* $\mathcal{M} \not\models_v^+ \psi$ *and* $\mathcal{M} \not\models_v^- \psi)$.
$\mathcal{M} \models_v^- \varphi \rightarrow \psi$ *iff* $\mathcal{M} \models_v^+ \varphi$ *and* $\mathcal{M} \models_v^- \psi$.

Logic of Paradox (LP): *Logic of Paradox* (LP) has been studied by Priest [21], which is one of the *paraconsistent logics* excluding (EFQ). As a paraconsistent logic, LP can be used to deal with various logical paradoxes and *Dialetheism* which is a philosophical position that admits some true contradictions. GC3 is taken as a sequent calculus of LP ([22]), and the truth table of LP can be obtained K_3's truth value **N** replaced with **B**.

The definition of the semantic relation for the implication of GC3 is obtained by replacing the implication in Definition 3.11 with the following definition.

$\mathcal{M} \models_v^+ \varphi \rightarrow \psi$ *iff* $\mathcal{M} \not\models_v^+ \varphi$ *or* $\mathcal{M} \not\models_v^- \psi$ or $(\mathcal{M} \models_v^+ \varphi$ *and* $\mathcal{M} \models_v^- \varphi$ *and* $\mathcal{M} \models_v^+ \psi$ *and* $\mathcal{M} \models_v^- \psi)$.
$\mathcal{M} \models_v^- \varphi \rightarrow \psi$ *iff* $\mathcal{M} \models_v^+ \varphi$ *and* $\mathcal{M} \models_v^- \psi$.

Belnap's four-valued logic: We have already seen the Belnap's four-valued logic. In addition, we defined the truth tables for $\sim$, $\wedge$, and $\vee$ as Table 3.4. The implication of B4 is defined with $\sim$ and $\vee$ and it does not hold for the rule of *modus ponens* because the *disjunctive syllogism* does not hold.

The aim of this chapter is to present many-valued semantics for the decision logic. There are three candidates of consequence relations for the enhancement in

Table 3.4 Truth tables for **B4**

~	N	F	T	B
	B	T	F	N

∧	N	F	T	B
N	N	F	N	F
F	F	F	F	F
T	N	F	T	B
B	F	F	B	B

∨	N	F	T	B
N	N	N	T	T
F	N	F	T	B
T	T	T	T	T
B	T	B	T	B

the decision logic. GC2, which was discussed above, is interpreted as strong Kleene three-valued logic. The value of a proposition is neither true nor false in GC2. In this case, the designated value of GC2 is defined as $\{\mathbf{T}, \mathbf{N}\}$.

GC3 is a paraconsistent logic, and its designated valued is defined as $\{\mathbf{T}, \mathbf{B}\}$. If the designated value of three-valued logic of GC3 is defined as $\{\mathbf{T}, \mathbf{B}\}$, therefore, it is possible to interpret the consequence relation by C3. GC4 is obtained from C4 based on four-valued semantics and interpreted as both paracomplete and paraconsistent.

Here, we present the extended version of many-valued logics with *weak negation* $\neg$. Weak negation represents the lack of truth. The assignment of weak negation is defined as follows:

$$\|\neg\varphi\|_s = \begin{cases} \mathbf{T} \text{ if } \|\varphi\|_s \neq \mathbf{T} \\ \mathbf{F} \text{ otherwise} \end{cases}$$

Weak implication is defined as follows:

$$A \rightarrow_w B =_{\text{def}} l\ /\!A \vee B$$

The assignment of weak implication is defined as follows:

$$\|A \rightarrow_w B\|^s = \begin{cases} \|B\|^s \text{ if } \|A\|^s \in \mathscr{D} \\ \mathbf{T} \text{ if } \|A\|^s \notin \mathscr{D} \end{cases}$$

We represent the truth tables for $\neg$ and $\rightarrow_w$ below (Table 3.5).

The semantic relations for weak negation are as follows:

$$\mathscr{M} \models_v^+ \neg\varphi \text{ iff } \mathscr{M} \not\models_v^+ \varphi,$$
$$\mathscr{M} \models_v^- \neg\varphi \text{ iff } \mathscr{M} \models_v^+ \varphi.$$

We try to extend many-valued logics with weak negation and weak implication. This regains some properties that some many-valued logics lack, such as the rule

Table 3.5 Truth tables for weak negation and weak implication

$\neg$	T	F	N	B
	F	T	T	F

$\rightarrow_w$	T	F	N	B
T	T	F	N	B
F	T	T	T	T
N	T	T	T	T
B	T	F	N	B

of *modus ponens* and the decision theorem. Obviously, L_3 recovers some properties that K_3 lacks and L_3's implication and weak implication has a close relationship.

Weak negation can represent the absence of truth. On the other hand, $\sim$ can serve as strong negation to express the verification of falsity. Note also that weak implication obeys the deduction theorem. This means that it can be regarded as a logical implication.

We can also interpret weak negation in terms of strong negation and weak implication:

$$\neg A =_{\text{def}} A \rightarrow_w \sim A$$

We define the sequent rules for $(\neg)$ and $(\rightarrow_w)$ as follows:

$$(\neg R)\frac{\Gamma, A \Rightarrow \Delta}{\Gamma \Rightarrow \neg A, \Delta} \quad (\neg L)\frac{\Gamma \Rightarrow A, \Delta}{\Gamma, \neg A \Rightarrow \Delta}$$

$$(\rightarrow_w R)\frac{A, \Gamma \Rightarrow \Delta, B}{\Gamma \Rightarrow \Delta, A \rightarrow_w B} \quad (\rightarrow_w L)\frac{B.\Gamma \Rightarrow \Delta \quad \Gamma \Rightarrow \Delta, A}{A \rightarrow_w B, \Gamma \Rightarrow \Delta}$$

GC2, GC3, and GC4 have additional rules of weak negation and weak implication, and we obtain $GC2^+$, $GC3^+$, and $GC4^+$. $GC2^+$ is the same as the extended Kleene logic EKL, that was proposed by Doherty [9] as the underlying three-valued logic for the non-monotonic logic and is provided with the deduction theorem.

$GC4^+$ is interpreted as both paracomplete and paraconsistent. This prevents the paradox of material implication of classical logic.

Here, it is observed that L_3 can be naturally interpreted in $GC2^+$. The Łukasiewicz implication can be defined as

$$A \supset B =_{\text{def}} (A \rightarrow_w B) \wedge (\sim B \rightarrow_w \sim A).$$

Next, we present the interpretation of weak negation for consequence relations C2, C3, and C4, which are interpreted as K_3, LP, and B4, respectively.

$$\|\neg A\|^{C2} = \begin{cases} \textbf{F} \text{ if } \|A\| = \textbf{T} \\ \textbf{T} \text{ if } otherwise \end{cases}$$

$$\|\neg A\|^{C3} = \begin{cases} \textbf{T} \text{ if } \|A\| = \textbf{F} \\ \textbf{F} \text{ if } otherwise \end{cases}$$

$$\|\neg A\|^{C4} = \begin{cases} \textbf{F} \text{ if } \|A\| = \textbf{T} \text{ or } \textbf{B} \\ \textbf{T} \text{ if } \|A\| = \textbf{F} \text{ or } \textbf{N} \end{cases}$$

We consider an application of weak negation for the interpretation of rough sets.

Example 3.3 Suppose the definition of a decision table is the same as Example 3.1.

$U = \{x_1, x_2, x_3, x_4, x_5, x_6, x_7, x_8\}$

Attribute: $C = \{c1, c2, c3, c4\}$

$c_1 = \{x_1, x_4, x_8\}$, $c_2 = \{x_2, x_5, x_7\}$, $c_3 = \{x_3\}$,
$c_4 = \{x_6\}$
$U/C = c_1 \cup c_2 \cup c_3 \cup c_4$
Any subset $X = \{x_3, x_6, x_8\}$
$POS_C(X) = c_3 \cup c_4 = \{x_3, x_6\}$
$BN_C(X) = c_1 = \{x_1, x_4, x_8\}$
$NEG_C(X) = c_2 = \{x_2, x_5, x_7\}$

The interpretation of the consequence relation C4 for weak negation in the decision table is defined as follows:

If $|C_{c3}|_S \subseteq POS_C(X)$ then $\neg\|C_{c3}\|^v = \textbf{F}$,
If $|C_{c2}|_S \subseteq NEG_C(X)$ then $\neg\|C_{c2}\|^v = \textbf{T}$,
If $|C_{c2}|_S \not\subseteq POS_C(X) \cup NEG_C(X)$ then $\neg\|C_{c2}\|^v = \textbf{T}$,
If $|C_{c1}|_S \subseteq BN_C(X)$ then $\neg\|C_{c1}\|^v = \textbf{F}$.

3.7 Soundness and Completeness

The soundness and completeness theorems are shown for the sequent system GB4$^+$, GB4$^+$ discussed above is interpreted as Belnap's four-valued logic B4 extended with weak negation and weak implication. The sequent calculus GB4$^+$ is defined as follows:

GB4$^+$:= {(ID), (*Weakening*), (*Cut*), $(\wedge R)$, $(\wedge L)$, $(\vee R)$, $(\vee L)$, $(\sim\sim R)$, $(\sim\sim L)$, $(\sim \wedge R)$, $(\sim \wedge L)$, $(\sim \vee R)$, $(\sim \vee L)$, $(\not R)$, $(\not L)$, $(\rightarrow_w R)$, $(\rightarrow_w L)$}

It is assumed that GB4$^+$ is the basic deduction system for decision logic obtained from GB4 with the ru;es for weak negation and weak implication. This prevents the paradox of material implication of classical logic.

The semantic relation of model $\mathcal{M}$ for GB4$^+$ is defined as follows:

Definition 3.21 The semantic relations of $\mathcal{M} \models^+_{GB4^+} \varphi$ and $\mathcal{M} \models^-_{GB4^+} \varphi$ are defined as follows:

$\mathcal{M} \models^+_{GB4^+} \top$,
$\mathcal{M} \models^-_{GB4^+} \bot$,
$\mathcal{M} \models^+_{GB4^+} p$ iff $p \in M^+$, where p is a propositional variable,
$\mathcal{M} \models^-_{GB4^+} p$ iff $p \in M^-$, where p is a propositional variable,
$\mathcal{M} \models^+_{GB4^+} \sim \varphi$ iff $\mathcal{M} \models^-_{GB4^+} \varphi$,
$\mathcal{M} \models^-_{GB4^+} \sim \varphi$ iff $\mathcal{M} \models^+_{GB4^+} \varphi$,
$\mathcal{M} \models^+_{GB4^+} \varphi \vee \psi$ iff $\mathcal{M} \models^+_{GB4^+} \varphi$ or $\mathcal{M} \models^+_{GB4^+} \psi$,
$\mathcal{M} \models^-_{GB4^+} \varphi \vee \psi$ iff $\mathcal{M} \models^-_{GB4^+} \varphi$ and $\mathcal{M} \models^-_{GB4^+} \psi$,
$\mathcal{M} \models^+_{GB4^+} \varphi \wedge \psi$ iff $\mathcal{M} \models^+_{GB4^+} \varphi$ and $\mathcal{M} \models^+_{GB4^+} \psi$,
$\mathcal{M} \models^-_{GB4^+} \varphi \wedge \psi$ iff $\mathcal{M} \models^-_{GB4^+} \varphi$ or $\mathcal{M} \models^-_{GB4^+} \psi$,
$\mathcal{M} \models^+_{GB4^+} \varphi \rightarrow \psi$ iff $\mathcal{M} \models^-_{GB4^+} \varphi$ or $\mathcal{M} \models^+_{GB4^+} \psi$,
$\mathcal{M} \models^-_{GB4^+} \varphi \rightarrow \psi$ iff $\mathcal{M} \models^+_{GB4^+} \varphi$ and $\mathcal{M} \models^-_{GB4^+} \psi$.
$\mathcal{M} \models^+_{GB4^+} \neg\varphi$ iff $\mathcal{M} \not\models^+_{GB4^+} \varphi$,
$\mathcal{M} \models^-_{GB4^+} \neg\varphi$ iff $\mathcal{M} \models^+_{GB4^+} \varphi$,
$\mathcal{M} \models^+_{GB4^+} \varphi \rightarrow_w \psi$ iff $\mathcal{M} \not\models^+_{GB4^+} \varphi$ or $\mathcal{M} \models^+_{GB4^+} \psi$,
$\mathcal{M} \models^-_{GB4^+} \varphi \rightarrow_w \psi$ iff ($\mathcal{M} \models^+_{GB4^+} \varphi$ or ($\mathcal{M} \models^+_{GB4^+} \varphi$ and $\mathcal{M} \models^-_{GB4^+} \varphi$)) and $\mathcal{M} \models^-_{GB4^+} \psi$.

Lemma 3.1 *The validity of the inference rules*

1. *The axioms of* GB4$^+$ *are valid.*
2. *For any inference rules of* GB4$^+$ *and any valuation s, if s satisfies all of the formulas of Pre, then s satisfies Cons.*

Proof (1) In GB4$^+$, the axiom (ID) preserves validity.

For (2), the proof for structural rules (weakening) and (cut), ($\not R$), ($\rightarrow_w L$), and ($\sim \wedge L$) will be provided.

$$(weakening)\frac{\Gamma \Rightarrow \Delta}{\Gamma' \Rightarrow \Delta'} \qquad (cut)\frac{\Gamma, A \Rightarrow \Delta \quad \Gamma \Rightarrow A, \Delta}{\Gamma \Rightarrow \Delta}$$

Suppose that $\models_{GB4^+} \Gamma \Rightarrow \Delta$. Then, clearly $V^4(\gamma) \neq \mathbf{T}$ for some $\gamma \in \Gamma'$ or $V^4(\delta) \neq \mathbf{F}$ for some $\delta \in \Delta'$.

Suppose that the principle is not correct, i.e. that (for some Γ, Δ, A) we have if (1) $\models_{GB4^+} \Gamma, A \Rightarrow \Delta$ and (2) $\models_{GB4^+} \Gamma \Rightarrow A, \Delta$ then (3) $\not\models_{GB4^+} \Gamma \Rightarrow \Delta$. Then by assumption (3) there is an interpretation V which assigns $\mathbf{T}$ to each formula in Γ, but assigns $\mathbf{F}$ to Δ.

If A occurs in Γ or Δ, then V cannot assign $\mathbf{F}$ to A. By assumption (1) and (2), V cannot assign $\mathbf{F}$ to Δ. This is a contradiction.

If A does not occur in Γ or Δ, then V is extended to V' by adding to its interpretation of A. Since V' agrees with V on the interpretation of all the formula in Γ and Δ, it will still be the case that V' assigns **T** to all formulae in Γ, and **F** to Δ. By the assumption (1) V' cannot assign **F** to A, and the assumption (2) V' cannot assign **T** to A since it assigns **T** to all in Γ but **F** to all in Δ. This is a contradiction.

$$(\neg R)\frac{\Gamma, A \Rightarrow \Delta}{\Gamma \Rightarrow \neg A, \Delta}$$

Suppose that $\models_{GB4^+} \Gamma, A \Rightarrow \Delta$. Then, either (4) $V^4(\gamma) \neq \mathbf{T}$ for some $\gamma \in \Gamma$ or $V^4(\delta) \neq \mathbf{F}$ for some $\delta \in \Delta$ or (5) $V^4(A) \neq \mathbf{T}$. If (4) holds, then clearly $\models_{GB4^+} \Gamma \Rightarrow \not{A}, \Delta$ *iff* $\models^-_{GB4^+} \Gamma$ *or* $\models^+_{GB4^+} \Delta, \neg A$. If (5) holds, then from the definition of $/$, it follows that $V^4(\neg A) = \mathbf{T}$ and then $\models_{GB4^+} \Gamma \Rightarrow \not{A}, \Delta$.

$$(\rightarrow_w L)\frac{B, \Gamma \Rightarrow \Delta \quad \Gamma \Rightarrow \Delta, A}{A \rightarrow_w B, \Gamma \Rightarrow \Delta}$$

Suppose that $\models_{GB4^+} B, \Gamma \Rightarrow \Delta$, and $\Gamma \Rightarrow \Delta, A$. Then, either (6) $V^4(\gamma) \neq \mathbf{T}$ for some $\gamma \in \Gamma$ or $V^4(\delta) \neq \mathbf{F}$ for some $\delta \in \Delta$ or (7) $V^4(B) \neq \mathbf{F}$ and $V^4(A) \neq \mathbf{T}$. If (6) holds, then clearly $\models_{GB4^+} A \rightarrow_w B, \Gamma \Rightarrow \Delta$. If (7) holds, then from the semantic relation of $\rightarrow_w$, it follows that $V^4(A \rightarrow_w B) \neq \mathbf{F}$ and again $\models_{GB4^+} A \rightarrow_w B, \Gamma \Rightarrow \Delta$.r

$$(\sim \wedge L)\frac{\sim A, \Gamma \Rightarrow \Delta \quad \sim B, \Gamma \Rightarrow \Delta}{\sim (A \wedge B), \Gamma \Rightarrow \Delta}$$

Suppose that $\models_{GB4^+} \sim A, \Gamma \Rightarrow \Delta$, and $\models_{GB4^+} \sim B, \Gamma \Rightarrow \Delta$. Then, either (8) $V^4(\gamma) \neq \mathbf{T}$ for some $\gamma \in \Gamma$ or $V^4(\delta) \neq \mathbf{F}$ for some $\delta \in \Delta$ or (9) $V^4(\sim A) \neq \mathbf{T}$ or $V^4(\sim B) \neq \mathbf{T}$. If (8) holds, then clearly $\models_{GB4^+} \sim (A \wedge B),\ \Gamma \Rightarrow \Delta$. If (9) holds, then from the definition of $\wedge$, it follows that $V^4(A \wedge B) \neq \mathbf{F}$, whence $V^4(\sim (A \wedge B)) = \mathbf{F}$, and again $\models_{GB4^+} \sim (A \wedge B),\ \Gamma \Rightarrow \Delta$.

Lemma 3.2 (Soundness of GB4$^+$) *If* $\vdash_{GB4^+} \Gamma \Rightarrow \Delta$ *is provable in GB4*$^+$*, then* $\models_{GB4^+} \Gamma \Rightarrow \Delta$.

Proof If the sequent $\Gamma \Rightarrow \Delta$ is an instance of axiom (ID), then $\Gamma \Rightarrow \Delta$ is valid in GB4$^+$. By induction on the depth of a derivation of $\Gamma \Rightarrow \Delta$ in GB4$^+$, it follows, by Lemma 3.1, that the sequent $\Gamma \Rightarrow \Delta$ is valid in GB4$^+$.

We are now in a position to prove the completeness of GB4$^+$. The proof below is similar to the Henkin proof described in Avron and Konikowska [4].

Theorem 3.8 (Completeness of GB4$^+$) *The sequent calculus GB4*$^+$ *is sound and complete for* $\models_{GB4^+}$.

Proof Let us denote the provability in GB4$^+$ by $\vdash_{GB4^+}$. For any sequent Σ over the language of GB4$^+$,

$$\vdash_{GB4^+} \Sigma \text{ if } \Sigma \text{ has a proof in GB4}^+.$$

We have to prove that, for any sequent Σ over the language of GB4$^+$,

$$\models_{GB4^+} \Sigma \text{ iff } \vdash_{GB4^+} \Sigma.$$

The backward implication, representing the soundness of the system, follows immediately from Lemma 3.2. To prove the forward implication *completeness*, we argue by contradiction. Suppose Σ is a sequent such that $\nvdash_{GB4^+} \Sigma$. We shall prove that $\not\models_{GB4^+} \Sigma$. Let us assume that the inclusion and union of sequents are defined componentwise, i.e.,

$(\Gamma' \Rightarrow \Delta') \subseteq (\Gamma'' \Rightarrow \Delta'')$ iff $\Gamma' \subseteq \Gamma''$ and $\Delta' \subseteq \Delta''$,
$(\Gamma' \Rightarrow \Delta') \cup (\Gamma'' \Rightarrow \Delta'') = \Gamma', \Gamma'' \Rightarrow \Delta', \Delta''$.

A sequent Σ_0 is called *saturated* if it is closed under all of the rules in GB4$^+$ applied backwards. More exactly, for any rule r in GB4$^+$ whose conclusion is contained in Σ_0, one of its premises must also be contained in Σ_0 (for a single premise rule, this means its only premise must be contained in Σ_0).

For example, if $\Sigma_0 = (\Gamma_0 \Rightarrow \Delta_0)$ is saturated and $(A \to B) \in \Delta_0$, then in view of the rules $(\to R)$, we must have both $\sim A \in \Delta$ and $B \in \Delta$. In turn, if $(A \to B) \in \Gamma_0$, then in view of the rule $(\to L)$, we must have either $\sim A \in \Gamma$ or $B \in \Gamma$.

Let $\Sigma = (\Gamma \Rightarrow \Delta)$ be any sequent. We shall first prove that Σ can be extended to a saturated sequent $\Sigma^* = (\Gamma^* \Rightarrow \Delta^*)$, which is not provable in GB4$^+$. If Σ is already saturated, we are done.

Otherwise, we start with the sequent Σ and expand it step by step by closing it under the subsequent rules of GB4$^+$ without losing the non-provability property. Specifically, we define a sequence $\Sigma_0, \Sigma_1, \Sigma_2, \ldots$ such that

1. $\Sigma_{i-1} \subseteq \Sigma_i$ for each $i \geq 1$,
2. Σ_i is not provable.

We take $\Sigma_0 = \Sigma_1 = \Sigma$; then, conditions 1 and 2 above are satisfied for $i = 1$. Assume that we have the constructed sequents $\Sigma_0, \Sigma_1, \ldots, \Sigma_k$ satisfying those conditions, and Σ_k is still not saturated. Then, there is a rule

$$r = \frac{\Pi_1 \ \cdots \ \Pi_l}{\Pi}$$

in GB4$^+$ such that $\Pi \subseteq \Sigma_k$ but $\Pi_i \nsubseteq \Sigma_k$ for $i = 1, \ldots, l$.

Since Σ_k is not provable, there must be an i such that $\Sigma_k \cup \Pi_i$ is not provable. Indeed, if $\Sigma_i \cup \Pi_i$ were provable for all i, $1 \leq i \leq l$, then we could deduce $\Sigma_k \cup \Pi$ from the provable sequents $\Sigma_k \cup \Pi_i$, $i = 1, \ldots, l$, using rule r, which in view of $\Sigma_k \cup \Pi = \Sigma_k$ would contradict the fact that Σ_k is not provable.

Thus, there is an i_0, $1 \leq i_0 \leq l$, such that $\Sigma_k \cup \Pi_{i_0}$ is not provable, and we take $\Sigma_{k+1} = \Sigma_k \cup \Pi_{i_0}$. Obviously, the sequents $\Sigma_0, \Sigma_1, \ldots, \Sigma_{k+1}$ satisfy conditions 1 and 2 above.

After a finite number n of such steps, we will have added all possible premises of the rules r in GB4$^+$ whose conclusions are contained in the original sequent Σ or its descendants in the constructed sequence, obtaining a saturated extension $\Sigma^* = \Sigma_n$ of Σ, which is not provable in GB4$^+$.

Thus, we have

- $\Sigma^* = (\Gamma^* \Rightarrow \Delta^*)$ is closed under the rules in GB4$^+$ applied backwards,
- $\Gamma \subseteq \Gamma^*, \Delta \subseteq \Delta^*$,
- $\nvdash_{GB4^+} \Sigma^*$.

We use Σ^* to define a counter-valuation for Σ, i.e., a legal valuation V^4 under the model of GB4$^+$ such that $\nvDash_{GB4^+} \Sigma$. For any propositional symbol $p \in P$ evaluated with the following valuation function, namely, we put:

$$V^4(p) = \begin{cases} \mathbf{T} \text{ if } p \in \Gamma \text{ and } p \notin \Delta, \\ \mathbf{F} \text{ if } \sim p \in \Gamma \text{ and } \sim p \notin \Delta, \\ \mathbf{B} \text{ if } \{p, \sim p\} \in \Gamma, \\ \mathbf{N} \text{ otherwise.} \end{cases} \tag{3.1}$$

For the valuation for the weak negation in GB4$^+$, define the following:

$$V^4(\neg p) = \begin{cases} \mathbf{T} \text{ if } V^4(p) \in \{\mathbf{F}, \mathbf{N}\}, \\ \mathbf{F} \text{ if } V^4(p) \in \{\mathbf{T}, \mathbf{B}\}. \end{cases} \tag{3.2}$$

For any A, B of the set of all well-formed formulas of GB4$^+$,

$$V^4(\sim A) = l \sim V^4(A). \tag{3.3}$$

$$V^4(A \rightarrow_w B) = \begin{cases} \mathbf{T} \text{ if } V^4(A) \in \{\mathbf{F}, \mathbf{N}\} \text{ or } V^4(B) = \mathbf{T}, \\ \mathbf{F} \text{ if } V^4(A) = \mathbf{T} \text{ and } V^4(B) = \mathbf{F}. \end{cases} \tag{3.4}$$

For $\wedge$ and $\vee$, we will write $\sqcap$ for the meet and $\sqcup$ for the join of $\leq$ order on **4**.

$$V^4(A \wedge B) = V^4(A) \sqcap V^4(B). \tag{3.5}$$

$$V^4(A \vee B) = V^4(A) \sqcup V^4(B). \tag{3.6}$$

It is easy to see that v defined as above is a well-defined mapping of the formulas of GC4$^+$ into **4**. Indeed, as Σ^* is not provable in GC4$^+$, then by (3.1), $v(p)$ is uniquely defined for any propositional symbol p, whence by (3.2), (3.3), $v(\varphi)$ is uniquely defined for any well-formed formula.

Moreover, by (3.2), (3.3), v is a legal interpretation of the language of GC4$^+$ under the interpretation of GC4$^+$, for the interpretations of $\sim$, $\rightarrow_w$ under v are compliant with the truth tables of those operations for this interpretation.

As Σ^* is an extension of Σ, in order to prove that $\nvDash_{GC4^+} \Sigma$, it suffices to prove that $\nvDash_{GC4^+} \Sigma^*$. We should prove for any well-formed formulas φ,

$$\models_{GC4^+} \gamma \text{ for any } \gamma \in \Gamma^*, \ \not\models_{GC4^+} \delta \text{ for any } \delta \in \Delta^*. \tag{3.7}$$

Equation (3.7) is proved by structural induction on the formulas in $S = \Gamma^* \cup \Delta^*$. We begin with literals in S, having the form of either p or $\sim p$, where $p \in P$. We have the following cases:

- $\varphi = p$. Then, by (3.1) and the fact that Γ^* and Δ^* are disjoint (for otherwise Σ^* would be provable), we have: $v(\varphi) \neq \mathbf{F}$ if $\varphi \in \Gamma^*$ and $v(\varphi) \neq \mathbf{T}$ if $\varphi \in \Delta^*$
- $\varphi = \sim p$. If $\varphi \in \Gamma^*$, then by (3.1), $v(p) \neq \mathbf{T}$, whence $v(\varphi) = \sim \mathbf{F} = \mathbf{T}$ by (3.3). In turn, if $\varphi \in \Delta^*$, then $\varphi \notin \Gamma^*$, whence $v(p) \neq \mathbf{F}$ and $v(\varphi) = \sim v(p) \neq \mathbf{T}$.
- $\varphi = \neg p$. If $\varphi \in \Gamma^*$, then by (3.1) $v(p) \neq \{\mathbf{T}, \mathbf{B}\}$, whence $v(\varphi) = \mathbf{T}$ by (3.2). In turn, if $\varphi \in \Delta^*$, then $\varphi \notin \Gamma^*$, whence $v(p) \neq \{\mathbf{F}, \mathbf{N}\}$ and $v(\varphi) =\sim v(p) = \mathbf{F}$.

Here, we define the rank ρ of formula φ by

$$\begin{aligned} \rho(p) &= 1, \\ \rho(\sim \varphi) &= \rho(\varphi) + 1, \\ \rho(\varphi \rightarrow \psi) &= \rho(\varphi) + \rho(\psi) + 1 \end{aligned}$$

Now we assume that the definition in (3.7) is satisfied for the formulas in S of rank up to n and suppose that $A, B \in S$ are at most of rank n. We prove that (3.7) holds for $\sim B$, $B \wedge C$ and $B \vee C$.

We begin with negation. Let $\varphi = \sim A$. As the case of $A = p \in P$ has already been considered, it remains to consider the following two cases:

- $A = \sim B$. Then, we have $\varphi = \sim\sim B$.
  - If $\varphi \in \Gamma^*$, then by rule $(\sim\sim L)$, we have $B \in \Gamma^*$, since Σ^* is a saturated sequent. Hence, by inductive assumption, $v(B) = \mathbf{T}$, and by (3.3), $v(\varphi) =\sim\sim \mathbf{T} = \mathbf{T}$.
  - In turn, if $\varphi \in \Delta^*$, then by rule $(\sim\sim R)$, we have $B \in \Delta^*$, whence by inductive assumption, $v(B) = \mathbf{F}$, and in consequence, $v(\varphi) = \sim\sim \mathbf{F} = \mathbf{F}$.
- $A = B \wedge C$. We again have two cases:
  - If $\varphi \in \Gamma^*$, then by rule $(\sim \wedge L)$, we have $\sim B, \sim C \in \Gamma^*$ since Σ^* is saturated. Hence, by inductive assumption, $v(B) \neq \mathbf{T}$ and $v(C) \neq \mathbf{T}$ (because $v(\sim B) \neq \mathbf{F}$ and $v(\sim C) \neq \mathbf{F}$). Thus, by the truth table $v(B \wedge C) \neq \mathbf{T}$; therefore, $v(\varphi) =\sim \mathbf{F} = \mathbf{T}$.
  - If $\varphi \in \Delta^*$, then by rule $(\sim \wedge R)$, we have either $\sim B \in \Delta^*$ or $\sim C \in \Delta^*$. By inductive assumption, this yields either $v(B) \neq \mathbf{T}$ or $v(C) \neq \mathbf{T}$. Thus, by the truth table, $v(B \wedge C) \neq \mathbf{T}$, whence $v(\varphi) =\sim \mathbf{T} = \mathbf{F}$.
- $A = B \vee C$. We again have two cases:
  - If $\varphi \in \Gamma^*$, then by rule $(\sim \vee L)$, we have $\sim B, \sim C \in \Gamma^*$ since Σ^* is saturated. Hence, by inductive assumption, $v(B) \neq \mathbf{T}$ and $v(C) \neq \mathbf{T}$ (because $v(\sim B) \neq \mathbf{F}$ and $v(\sim C) \neq \mathbf{F}$). Thus, $v(B \vee C) \neq \mathbf{T}$, and $v(\varphi) =\sim \mathbf{F} = \mathbf{T}$.

– If $\varphi \in \Delta^*$, then by rule $(\sim \vee R)$ we have either $\sim B \in \Delta^*$ or $\sim C \in \Delta^*$. By inductive assumption, this yields either $v(B) \neq \mathbf{F}$ or $v(C) \neq \mathbf{F}$. Thus, $v(B \vee C) \neq \mathbf{F}$, whence $v(\varphi) \neq \mathbf{T} = \mathbf{F}$.

It remains to consider implication. Let $\varphi = A \rightarrow_w B$. We have the following two cases:

- $\varphi \in \Gamma^*$. Then, as Σ^* is saturated, by rule $(\rightarrow_w L)$, we have either $A \in \Delta^*$ or $B \in \Gamma^*$. In view of (3.1) and (3.3), and the fact that $\varphi \notin \Delta^*$, this yields either $v(A) \in \{\mathbf{F}, \mathbf{N}\}$ or $v(B) \in \{\mathbf{T}, \mathbf{B}\}$. Thus $v(A \rightarrow_w B) \neq \mathbf{F}$, and $v(\varphi) = \mathbf{T}$.
- $\varphi \in \Delta^*$. Then, as Σ^* is saturated, by rules $(\rightarrow_w R)$ we have $A \in \Gamma^*$ and $B \in \Delta^*$. In view of (3.1) and (3.3), and the fact that $\varphi \notin \Gamma^*$, this yields $v(A) \in \{\mathbf{T}, \mathbf{B}\}$ and $v(B) \in \{\mathbf{F}, \mathbf{N}\}$, thus, $v(A \rightarrow_w B) \neq \mathbf{T}$, and $v(\varphi) = \mathbf{F}$.

Thus, (3.7) holds, and $\models_{GC4^+} \Sigma$, which ends the completeness proof.

GC4$^+$ may be one candidate for the extended version of decision logic that is needed to handle uncertain information and be tolerant to inconsistency.

3.8 Conclusion and Future Work

In this chapter, we propose an extension of the decision logic of rough sets to handle uncertainty, ambiguity and inconsistent states in information systems based on rough sets.

We investigate some properties of information system based on rough sets and define some characteristics of a certain relationship for the interpretation of truth values.

We obtain some observations for a relationship between the interpretation with four-valued truth values and the regions defined with rough sets. To handle these characteristics we have introduced partial semantics with consequence relations for the axiomatization with many-valued logics and proposed a unified formulation of the decision logic of rough sets and many-valued logics.

We also extend the language of many-valued logics with weak negation to enable the deduction theorem or the rule of modus ponens. We have shown that the system GC4$^+$ is sound and complete with Belnap's four-valued semantics.

In future work, the extension of language should be investigated, e.g., an operator to handle the granularity of objects or the uncertainty of a proposition, which is related to some kind of modal operators to recognize the crispness of objects.

In this chapter, we introduce the rules of weak negation and weak implication to extend many-valued logics. They can enhance many-valued logics for handling a deduction system more usefully.

To grasp the information state represented with information in detail, another extension of language should be investigated, such as modal type operators in a paraconsistent version of Łukasiewicz logic J3; see Epstein [10].

Furthermore, we need to investigate another version of decision logics based on an extended version of rough set theories, e.g., the variable precision rough set (VPRS) [26].

VPRS models are an extension of standard (Pawlak-type) rough set model, which enables us to treat probabilistic or inconsistent information in the framework of rough sets. The fact implies that we could have several non-classical versions of decision logics.

As said above, there are many topics to be explored. By these further investigations, a much more useful version of extended decision logic is expected for practical applications and actual data analytics.

References

1. Akama, S., Murai, T., Kudo, Y.: Reasoning with Rough Sets. Springer, Heidelberg (2018)
2. Akama, S., Nakayama, Y.: Consequence relations in DRT. In: Proc. of the 15th International Conference on Computational Linguistics (COLING 1994) 2, pp. 1114–1117 (1994)
3. Anderson, A., Belnap, N.: Entailment: The Logic of Relevance and Necessity I. Princeton University Press, Princeton (1976)
4. Avron, A., Konikowska, B.: Rough sets and 3-valued logics. Studia Logica **90**, 69–92 (2008)
5. Belnap, N.D.: A useful four-valued logic. In: Dunn, J.M., Epstein, G. (eds.) Modern Uses of Multi-Valued Logic, pp. 8–37. Reidel, Dordrecht (1977)
6. Belnap, N.D.: How a computer should think. In: Ryle, G. (ed.) Contemporary Aspects of Philosophy, pp. 30–55. Oriel Press (1977)
7. Ciucci, D., Dubois, D.: Three-valued logics, uncertainty management and rough sets. In: Transactions on Rough Sets XVII. Lecture Notes in Computer Science Book Series (LNCS), 8375, pp. 1–32. Springer, Berlin (2001)
8. Degauquier, V.: Partial and paraconsistent three-valued logics. Logic Logical Philos. **25**, 143–171 (2016)
9. Doherty, P.: NM3–A three-valued cumulative non-monotonic formalism. In: Eijck, van. (ed.) Proc. of European Workshop on Logics in Artificial Intelligence (JELIA), pp 196–211. Sprnger, Heidelberg (1990)
10. Epstein, R.L.: The Semantic Foundations of Logic. Springer, Heidelberg (1990)
11. Fan, T.-F., Hu, W.-C., Liau, C.-J.: Decision logics for knowledge representation in data mining. In: Proc. of the 25th Annual International Computer Software and Applications Conference (COMPSAC), pp. 626–631 (2001)
12. Kleene, S.: Introduction to Metamathematics. North-Holland, Amsterdam (1952)
13. Lin, Y., Qing, L.: A logical method of formalization for granular computing. In: Proc. of the IEEE International Conference on Granular Computing (GRC 2007), pp. 22–27 (2007)
14. Łukasiewicz, J.: On 3-valued logic, 1920. In: McCall, S. (ed.) Polish Logic, pp. 16–18. Oxford University Press, Oxford (1967)
15. Łukasiewicz, J.: Many-valued systems of propositional logic, 1930. In: McCall, S. (ed.) Polish Logic. Oxford University Press, Oxford (1967)
16. Muskens, R.: On partial and paraconsistent logics. Notre Dame J. Formal Logic **40**, 352–374 (1999)
17. Nakayama, Y., Akama, S., Murai, T.: Deduction system for decision logic based on Partial semantics. In: Proc. of the 11th International Conference on Advances in Semantic Processing, pp. 8–11 (2017)
18. Nakayama, Y., Akama, S., Murai, T.: Deduction system for decision logic based on many-valued logics. Int. J. Adv. Intell. Syst. **11**, 115–126 (2018)

19. Pawlak, Z.: Rough sets. Int. J. Comput. Inf. Sci. **11**, 341–356 (1982)
20. Pawlak, Z.: Rough Sets: Theoretical Aspects of Reasoning about Data. Kluwer, Dordrecht (1991)
21. Priest, G.: The logic of paradox. J. Philos. Logic **8**, 219–241 (1979)
22. Priest, G.: An Introduction to Non-Classical Logic, From If to Is, 2nd edn. Cambridge University Press, Cambridge (2008)
23. Smullyan, R.: First-Order Logic. Springer, Berlin (1968)
24. Urquhart, A.: Basic many-valued logic. In: Gabbay, G., Guenthner, F. (eds.) Handbook of Philosophical Logic, 2, pp. 249–295. Springer, Heidelberg (2001)
25. van Benthem, J.: Partiality and nonmonotonicity in classical logic. Logique et Analyse **29**, 225–247 (1986)
26. Ziarko, W.: Variable precision rough set model. J. Comput. Syst. Sci. **46**, 39–59 (1993)

Chapter 4
Tableaux Calculi for Many-Valued Logics

Abstract In Chap. 4, we present another deduction system based on tableaux calculus for four-valued logic and introduce the analytic tableaux as a basis for an automated deductive system.

4.1 Introduction

This chapter aims to present many-valued semantics and tableaux calculus as a deduction basis for the decision logic. It is an extended version of Nakayama et al. [20]. *Rough set theory* was studied by Pawlak for handling rough and coarse information [21, 22]. In applying rough set theory, decision logic was proposed by Pawlak for interpreting information extracted from data tables.

However, decision logic is based on the premise that data tables are *consistent* and defined with the classical two-valued logic. It is known that classical logic is not adequate for reasoning with undefined and inconsistent information.

To handle incomplete information, we introduce Belnap's four-valued logic. In Belnap [6], he claimed that both incomplete and inconsistent information should be expressed in a database. Therefore, four-valued logic is suitable to handle inconsistency on decision tables and it can serve as a deductive basis for decision logic.

At the same time, to define four-valued interpretation to inconsistent data table, we introduce *variable precision rough set* (for short, VPRS). *VPRS model* was proposed by Ziarko [29] to grasp inconsistent state of data table.

VPRS is an extension of Pawlak's rough set theory which provides a theoretical basis to treat probabilistic or inconsistent information in the framework of rough sets.

Belnap's four-valued logic serves to compensate VPRS as following meanings. Four-valued semantics can handle both inconsistent information and incomplete information represented with VPRS, and four-valued logic can provide a proof theoretic formulation to serve as a deduction basis for VPRS.

S. Akama et al., *Epistemic Situation Calculus Based on Granular Computing*,
Intelligent Systems Reference Library 239,
https://doi.org/10.1007/978-3-031-28551-6_4

For example, VPRS can represent objects in a boundary region as undefined or inconsistent. Additionally, VPRS can also represent objects both *true* and *false* at the same time by an intersection of $\beta-$positive region and $\beta-$negative region. Four-valued logic can handle in these cases and also provide deduction basis for decision tables.

As for a proof system in this study, we adopt tableaux calculi as a basis for deductive system and show the Henkin type soundness and completeness proof for many-valued tableaux calculus. The deductive system of decision logic has been studied from the granule computing perspective, and in Fan et al. [12], an extension of decision logic was proposed for handling uncertain data tables by fuzzy and probabilistic methods.

In Lin and Qing [14], a natural deduction system based on classical logic was proposed for decision logic in granule computing. In Avron and Konikowska [5], Gentzen-type three-valued sequent calculus was proposed for rough set theory based on non-deterministic matrices for semantic interpretation.

Gentzen type axiomatization of three-valued logics based on partial semantics for decision logic is proposed in Nakayama et al. [18]. Vitoria et al. [28] propose set-theoretical operations on four-valued sets for rough sets. The reasoning with rough sets is comprehensively studied in Akama et al. [2].

The chapter is organized as follows. In Sect. 4.2, an overview of rough sets, decision logic and Belnap's four-valued logic are presented. In Sect. 4.3, the relationship between decision logic and three-valued semantics based on partial semantics is discussed. A semantic interpretation between rough sets and a four-valued logic is also investigated, and we propose systematization with tableaux calculi for a deduction system for four-valued semantics.

In Sect. 4.4, a Henkin-style completeness proof is provided for four-valued semantics extended with weak negation. Finally, in Sect. 4.5, a summary of the study and possible directions for future work are provided.

4.2 Backgrounds

4.2.1 Rough Set and Decision Logic

Rough set theory, proposed by Pawlak [21], provides a theoretical basis of sets based on approximation concepts. A rough set can be seen as an approximation of a set. It is denoted by a pair of sets, called the lower and upper approximation of the set.

Rough sets are used for imprecise data handling. For the upper and lower approximations, any subset X of U can be in any of three states, according to the membership relation of objects in U.

If the positive and negative regions on a rough set are considered to correspond to the truth value of a logical form, then the boundary region corresponds to ambiguity in deciding truth or falsity. Thus, it is plausible to adopt three-valued and four-valued logics for the basis for rough sets.

Rough set theory is outlined below. Let U be a non-empty finite set, called a universe of objects. If R is an equivalence relation on U, then U/R denotes the family of all equivalence classes of R, and the pair (U, R) is called a Pawlak approximation space, which is defined as follows:

Definition 4.1 Let R be an equivalence relation of the approximation space $S = (U, R)$, and X any subset of U. Then, the lower and upper approximations of X for R are defined as follows:

$$\underline{R}X = \bigcup\{Y \in U/R \mid Y \subseteq X\} = \{x \in U \mid [x]_R \subseteq X\},$$
$$\overline{R}X = \bigcup\{Y \in U/R \mid Y \cap X \neq 0\} = \{x \in U \mid [x]_R \cap X \neq \emptyset\},$$

Definition 4.2 If $S = (U, R)$ and $X \subseteq U$, then the *R-positive*, *R-negative*, and *R-boundary* regions of X with respect to R are defined respectively as follows:

$$POS_R(X) = \underline{R}X,$$
$$NEG_R(X) = U - \overline{R}X,$$
$$BN_R(X) = \overline{R}X - \underline{R}X.$$

Objects included in R-boundary are interpreted as *undefined* and *inconsistent.* In general, targets of a decision logic are described by table-style format called *information tables.*

Information table used by Pawlak [22] is defined by $T = (U, A, C, D)$, where U is a finite and non-empty set of objects, A is a finite and non-empty set of attributes. C and D are subsets of a set of attributes A and C $(D \subseteq A)$, and it is assumed that C is a conditional attribute and D a decision attribute.

Definition 4.3 The set of formulas of the decision logic language *DL* is the smallest set satisfying the following conditions:

1. (a, v), or in short a_v, is an atomic formula of *DL*, where the set of attribute constants is defined as $a \in A$ and the set of attribute value constants is $v \in V = \bigcup V_a$.
2. If φ *and* ψ are formulas of the *DL*, then $\sim \varphi$, $\varphi \wedge \psi$, $\varphi \vee \psi$, $\varphi \rightarrow \psi$, and $\varphi \equiv \psi$ are formulas.

The interpretation of DL is performed using the universe U in the Knowledge Representation System (*KR-system*) $K = (U, A)$ and the assignment function s, mapping from U to objects of formulas defined as follows:

$$|\varphi|_S = \{x \in U : x \models_S \varphi\}.$$

Formulas of *DL* are interpreted as subsets of objects consisting of a value v and an attribute a.

The semantic relations of compound formulas are recursively defined as follows:

$x \models_S a(x, v)$ iff $a(x) = v$,
$x \models_S \sim \varphi$ iff $x \not\models_S \varphi$,
$x \models_S \varphi \vee \psi$ iff $x \models_S \varphi$ or $x \models_S \psi$,

$x \models_S \varphi \wedge \psi$ iff $x \models_S \varphi$ and $S \models_S \psi$,
$x \models_S \varphi \rightarrow \psi$ iff $x \models_S \sim \varphi \vee \psi$,
$x \models_S \varphi \equiv \psi$ iff $x \models_S \varphi \rightarrow \psi$ and $s \models_S \psi \rightarrow \varphi$.

Let φ be an atomic formula of D, $R \in C \cup D$ an equivalence relation, and X any subset of U, and a valuation v of propositional variables.

$$\|\varphi\|^v = \begin{cases} \mathbf{t} \text{ if } |\varphi|_S \subseteq POS_R(U/X) \\ \mathbf{f} \text{ if } |\varphi|_S \subseteq NEG_R(U/X) \end{cases}.$$

This shows that decision logic is based on bivalent logic.

4.2.2 *Variable Precision Rough Set*

Variable precision rough set models (for short, VPRS) proposed by Ziarko [29] is one extension of Pawlak's rough set theory which provides a theoretical basis to treat probabilistic or inconsistent information in the framework of rough sets.

VPRS is based on the majority inclusion relation. Let $X, Y \subseteq U$ be any subsets of U. The majority inclusion relation is defined by the following measure $c(X, Y)$ of the relative degree of misclassification of X with respect to Y,

$$c(X, Y) =_{\text{def}} \begin{cases} 1 - \dfrac{|X \cap Y|}{|X|}, & \text{if } X \neq \emptyset, \\ 0, & \text{otherwise.} \end{cases}$$

where $|X|$ represents the cardinality of the set X. It is easy to confirm that $X \subseteq Y$ holds, if and only if $cd(X, Y) = 0$.

Formally, the majority inclusion relation $\subseteq$ with a fixed precision $b \in [0, 0.5)$ is defined using the relative degree of misclassification as follows:

$$X \overset{\beta}{\subseteq} Y \text{ iff } c(X, Y) \leq \beta,$$

where the precision b provides the limit of permissible misclassification.

Let $X \subseteq U$ be any set of objects, R be an indiscernibility relation on U, and a degree $\beta \in [0, 0.5)$ be a precision. he $\beta -$ *lower approximation* $\underline{R}_\beta(X)$ of X and the $\beta -$ *upper approximation* $\overline{R}_\beta(X)$ of X by R are defined as follows, respectively:

$$\begin{aligned} \underline{R}_\beta(X) &=_{\text{def}} \left\{x \in U \mid c([x]_R, X) \leq \beta\right\}, \\ \overline{R}_\beta(X) &=_{\text{def}} \left\{x \in U \mid c([x]_R, X) \leq 1 - \beta\right\}. \end{aligned}$$

As mentioned previously, the precision β represents the threshold degree of misclassification of elements in the equivalence class $[x]_R$ to the set X. Thus, in VPRS, misclassification of elements is allowed if the ratio of misclassification is less than β. Note that the $\beta -$ lower and $\beta -$ upper approximations with $\beta = 0$ correspond to Pawlak's lower and upper approximations.

4.2.3 Belnap's Four-Valued Logic

Belnap [7] first claimed that an inference mechanism for database should employ a certain four-valued logic. The important point in Belnap's system is that we should deal with both incomplete and inconsistent information in databases.

To represent such information, we need a four-valued logic, since classical logic is not appropriate to the task. Belnap's four-valued semantics can be in fact viewed as an intuitive description of internal states of a computer.

In Belnap's four-valued logic **B4**, four kinds of truth-values are used from the set **4** = {**T**, **F**, **N**, **B**}. These truth-values can be interpreted in the context of a computer, namely **T** means just told True, **F** means just told False, **N** means told neither True nor False, and **B** means told both True and False. Intuitively, **N** can be interpreted as undefined, and **B** as overdefined, respectively.

Belnap outlined a semantics for **B4** using the logical connectives. Belnap's semantics uses a notion of *set-ups* mapping atomic formulas into **4**. A set-up can then be extended for any formula in **B4** in the following way:

$$s(A \ \& \ B) = s(A) \ \& \ s(B),$$
$$s(A \ \vee \ B) = s(A) \ \vee \ s(B),$$
$$s(\sim A) = \sim s(A).$$

Belnap also defined a concept of entailments in **B4**. We say that A entails B just in case for each assignment of one of the four values to variables, the value of A does not exceed the value of B in B4, i.e. $s(A) \leq s(B)$ for each set-up s. Here, $\leq$ is defined as: $\mathbf{F} \leq \mathbf{B}, \mathbf{F} \leq \mathbf{N}, \mathbf{B} \leq \mathbf{T}, \mathbf{N} \leq \mathbf{T}$. Belnap's four-valued logic in fact coincides with the system of *tautological entailments* due to Anderson and Belnap [4].

Belnap's logic **B4** is one of paraconsistent logics capable of tolerating contradictions. Belnap also studied implications and quantifiers in **B4** in connection with *question-answering systems*.

Definition 4.4 defines a partial model for decision logic DL.

Definition 4.4 (Partial Model) A partial model for decision logic language DL is a tuple $\mathscr{M} = (\mathscr{T}, \mathscr{D}, \mathscr{O})$, where

- $\mathscr{T}$ is a non-empty set of truth-values defined as $\mathbf{4} = \{\mathbf{T}, \mathbf{F}, \mathbf{N}, \mathbf{B}\}$.
- $\emptyset \subset \mathscr{D} \subseteq \mathscr{T}$ is the set of designated values defined as $\mathscr{D} = \{\mathbf{T}, \mathbf{B}\}$.
- For every *n-ary connective* $\diamond$ of DL, $\mathscr{O}$ includes a corresponding n-ary function $\tilde{\diamond}$ from $\mathscr{T}^n$ to **4**.

4.2.4 Analytic Tableaux

For a deduction system based on a four-valued logic, we focus on the partial semantics. In classical logic, partial semantics is closely related to the interpretation of the Beth tableau [25]. Van Benthem [27] suggested the relationship for the consequence relation to a tableaux calculus.

We begin by recalling the basic idea of the Beth tableau. The Beth tableaux proves $X \rightarrow Y$ by constructing a counterexample of $X \,\&\, \sim Y$. The Beth tableaux has several partial features. For instance, there may be counterexamples even if a branch remains open. This insight led van Benthem [27] to work out partial semantics for classical logic.

In the Beth tableau, it is assumed that V is a partial valuation function that assigns the values 0 or 1 to an atomic formula p. We can then set $V(p) = 1$ for p on the left-hand side and $V(p) = 0$ for p on the right-hand side in an open branch of tableaux.

We describe analytic tableaux which is a variant of the "semantic tableaux" according to Smullyan [25]. Our present formulation is based on Akama [1]. The basic idea of tableaux calculi derives from Gentzen [13]. We begin by noting that under any interpretation the following eight inference rules hold for any formulas X, Y:

1) a) If $\sim X$ is true, then X is false.
   b) If $\sim X$ is false, then X is true.
2) a) If a conjunction $X \wedge Y$ is true, then X, Y are both true.
   b) If a conjunction $X \wedge Y$ is false, then either X is false or Y is false.
3) a) If a disjunction $X \vee Y$ is true, then either X is true or Y is true.
   b) If a disjunction $X \vee Y$ is false, then both X, Y are false.
4) a) If $X \rightarrow Y$ is true, then either X is false or Y is true.
   b) If $X \rightarrow Y$ is false, then X is true and Y is false.

These eight facts provide the basis of the tableaux method. At this stage it will prove useful to introduce the symbols "T", "F" to the object language, and define a *signed formula* as an expression TX or FX, where X is a (unsigned) formula. We read "TX" as "X is true" and "FX" as "X is false".

Definition 4.5 Under any interpretation, a signed formula TX is called true if X is true, and false if X is false. And a signed formula FX is called true if X is false, and false if X is true.

Thus the truth value of TX is the same as that of X; the truth value of FX is the same as that of $\sim X$. By the conjugate of a signed formula we mean the result of changing "T" to "F" or "F" to "T" (thus the conjugate of TX is FX; the conjugate of FX is TX).

We now state all the rules in schematic form; explanations immediately follow. For each logical connective there are two rules; one for a formula preceded by "T", the other for a formula preceded by "F":

$$\text{(1)}\quad \frac{T\sim X}{FX} \qquad \frac{F\sim X}{TX}$$

$$\text{(2)}\quad \frac{T(X\wedge Y)}{\begin{matrix}TX\\TY\end{matrix}} \quad (\quad \frac{F(X\wedge Y)}{FX\mid FY}$$

$$\text{(3)}\quad \frac{T(X\vee Y)}{T(X\mid Y)} \qquad \frac{F(X\vee Y)}{\begin{matrix}FX\\FY\end{matrix}}$$

$$\text{(4)}\quad \frac{T(X\rightarrow Y)}{FX\mid TY} \qquad \frac{F(X\rightarrow Y)}{\begin{matrix}TX\\FY\end{matrix}}$$

Rule (1) means that from $T \sim X$ we can directly infer FX (in the sense that we can subjoin FX to any branch passing through $T \sim X$) and that from $F \sim X$ we can directly infer TX. Rule (2) means that $T(X \wedge Y)$ directly yields both TX, TY, whereas $F(X \wedge Y)$ branches into FX, FY. Rules (3) and (4) can now be understood analogously.

Signed formulas, other than signed variables, are of two types;

(A) those which have direct consequences (viz. $F \sim X$, $T \sim X$, $T(X \wedge Y)$, $F(X \vee Y)$, $F(X \rightarrow Y)$;
(B) those which branch (viz. $F(X \wedge Y)$, $T(X \vee Y)$, $T(X \rightarrow Y)$.

It is practically desirable in constructing a tableau, that when a line of type (A) appears on the tableau, we simultaneously subjoin its consequences to all branches which pass through that line.

Then that line need never be used again. And in using a line of type (B), we divide all branches which pass through that line into sub-branches, and the line need never be used again.

If we construct a tableaux in the above manner, it is not difficult to see, that after a finite number of steps we must reach a point where .every line has been used (except of course, for signed variables, which are never used at all to create new lines). At this point our tableaux is complete (in a precise sense which we will subsequently define).

One way to complete a tableaux is to work systematically downwards i.e. never to use a line until all lines above it (on the same branch) have been used. Instead of this procedure, however, it turns out to be more efficient to give priority to lines of type (A)—i.e. to use up all such lines at hand before using those of type (B).

In this way, one will omit repeating the same formula on different branches; rather it will have only one occurrence above all those branch points.

As an example of both procedures, let us prove the formula $[p \rightarrow (q \rightarrow r)] \rightarrow [(P \rightarrow q) \rightarrow (p \rightarrow r)]$. tableaux works systematically downward; the second uses the second suggestion. For the convenience of the reader, we put to the right of each line the number of the line from which it was inferred.

It is apparent that Table 4.2 is quicker to construct than Table 4.1, involving only 13 rather than 23 lines.

Table 4.1 First Tableaux

$(1)F[p \to (q \to r)] \to [(p \to q) \to (p \to r)]$					
$(2)Tp \to (q \to r)$ (1)					
$(3)F(p \to q) \to (p \to r)$ (1)					
(4) Fp (2)		(5) $T(q \to r)$ (2)			
(6) $T(p \to q)$ (3)		(8) $T(p \to q)$ (3)			
(7) $F(p \to r)$ (3)		(9) $F(p \to r)$ (3)			
(10) Fp (6)	(11) Tq (6)	(14) Fq (5)		(15) Tr (5)	
(12) Tp (7)	(13) Tp (7)	(16) Fp (8)	(17) Tq (8)	(18) Fp (8)	(19) Tq (8)
X	X	(20) Tp (9)	X	(21) Tp (9)	(22) Tp (9)
		X		X	(23) Fr (9)
					X

Table 4.2 Second Tableaux

$(1)F[p \to (q \to r)] \to [(p \to q) \to (p \to r)]$			
$(2)Tp \to (q \to r)$ (1)			
$(3)F(p \to q) \to (p \to r)$ (1)			
$(4)T(p \to q)$ (3)			
$(5)F(p \to r)$ (3)			
$(6)T(p)$ (5)			
$(7)F(r)$ (5)			
(8) Fp (2)	(9) $T(q \to r)$ (2)		
X	(10) $F(p)$ (4)	(10) $T(q)$ (4)	
	X	(14) Fq (5)	(15) Tr (5)
		X	X

The method of analytic tableaux can also be used to show that a given formula is a truth functional consequence of a given finite set of formulas. Suppose we wish to show that $X \to Z$ is a truth-functional consequence of the two formulas $X \to Y$, $Y \to Z$.

We could, of course, simply show that $[(X \to Y) \wedge (Y \to Z)] \to (X \to Z)$ is a tautology. Alternatively, we can construct a tableaux starting with

$T(X \to Y)$,
$T(Y \to Z)$,
$F(X \to Z)$

and show that all branches close.

In general, to show that Y is truth-functionally implied by $X_1, \ldots, X_n$, we can construct either a closed analytic tableaux starting with $F(X_1 \wedge \cdots \wedge X_n) \to Y$, or one starting with

$$\begin{array}{c} TX_1 \\ \vdots \\ TX_n \\ FY \end{array}$$

Our use of the letters "T" and "F", though perhaps heuristically useful, is theoretically quite dispensable–simply delete every "T" and substitute "$\sim$" for "F". (In which case, incidentally, the first half of Rule 1) becomes superfluous.) The rules then become:

$$(1)\quad \frac{\sim\sim X}{X}$$

$$(2)\quad \frac{X \wedge Y}{\begin{array}{c} X \\ Y \end{array}} \qquad \frac{\sim (X \wedge Y)}{\sim X \mid\sim Y}$$

$$(3)\quad \frac{X \vee Y}{X \mid Y} \qquad \frac{\sim (X \vee Y)}{\begin{array}{c} \sim X \\ \sim Y \end{array}}$$

$$(4)\quad \frac{X \rightarrow Y}{\sim X \mid Y} \qquad \frac{\sim (X \rightarrow Y)}{\begin{array}{c} X \\ \sim Y \end{array}}$$

In working with tableaux which use unsigned formulas, "closing" a branch naturally means terminating the branch with a cross, as soon as two formulas appear, one of which is the negation of the other. A tableaux is called *closed* if every branch is closed.

Definition 4.6 (Closed branch) A closed branch is a branch which contains a formula and its negation (conjugate).

Definition 4.7 (Open branch) An open branch is a branch which is not closed.

Definition 4.8 (Closed tableaux) A tableaux is closed if all its branches are closed.

By a tableaux for a formula X, we mean a tableaux which starts with X. If we wish to prove a formula X to be a tautology, we construct a tableaux not for the formula X, but for its negation $\sim X$.

It will save us considerable repetition of essentially the same arguments in our subsequent development if we use the following unified notation which we introduced in [25].

We use the letter "α" to stand for any signed formula of type A-i.e. of one of the five forms $T(X \wedge Y)$, $F(X \vee Y)$, $F(X \rightarrow Y)$, $T \sim X$, $F \sim X$.For every such formula α, we define the two formulas α_1 and α_2 as follows:

$\alpha = T(X \wedge Y)$, then $\alpha_1 = TX$ and $\alpha_2 = TY$.
$\alpha = F(X \vee Y)$, then $\alpha_1 = FX$ and $\alpha_2 = FY$.
$\alpha = F(X \rightarrow Y)$, then $\alpha_1 = TX$ and $\alpha_2 = FY$.
$\alpha = T \sim X$, then $\alpha_1 = FX$ and $\alpha_2 = FX$.
$\alpha = F \sim X$, then $\alpha_1 = TX$ and $\alpha_2 = TX$.

For perspicuity, we summarize these definitions in the following table:

α	α_1	α_2
$T(X \wedge Y)$	TX	TY
$F(X \vee Y)$	FX	FY
$F(X \rightarrow Y)$	TX	FY
$T \sim X$	FX	FX
$F \sim X$	TX	TX

We note that in any interpretation, α is true iff α_1, α_2 are both true. Accordingly, we shall also refer to an (α as a formula of *conjunctive* type.

We use "β" to stand for any signed formula of type B-i.e. one of the three forms $F(X \wedge Y)$, $T(X \vee Y)$, $T(X \rightarrow Y)$. For every such formula β, we define the two formulas β_l, β_2 as in the following table:

β	β_1	β_2
$F(X \wedge Y)$	FX	FY
$T(X \vee Y)$	TX	TY
$T(X \rightarrow Y)$	FX	TY

In any interpretation, β is true iff at least one of the pair β_1, β_2 is true. Accordingly, we shall refer to any β-type formula as a formula of *disjunctive* type.

We shall sometimes refer to α_1 as the first component of α_1 and α_2 as the second component of α. Similarly, for β.

By the degree of a signed formula TX or FX we mean the degree of X. We note that α_1, α_2 are each of lower degree than α, and β_1, β_2 are each of lower degree than β. Signed variables, of course, are of degree O.

We might also employ an α, β classification of unsigned formulas in an analogous manner, simply delete all "T", and replace "F" by "$\sim$". The tables would be as shown in (Table 4.3).

Let us now note that whether we work with signed or unsigned formulas, all our tableaux rules can be succinctly lumped into the following two:

$$\text{Rule } A \ \frac{\alpha}{\begin{array}{c}\alpha_1\\ \alpha_2\end{array}} \qquad\qquad \text{Rule } B \ \frac{\beta}{\beta_1 \mid \beta_2}$$

Table 4.3 Rules for unsigned formulas

α	α_1	α_2
$X \wedge Y$	X	Y
$\sim (X \vee Y)$	$\sim X$	$\sim Y$
$\sim (X \rightarrow Y)$	X	$\sim Y$
$\sim\sim X$	X	X

β	β_1	β_2
$\sim (X \wedge Y)$	$\sim X$	$\sim Y$
$X \vee Y$	X	Y
$X \rightarrow Y$	$\sim X$	Y

The operation of conjugation obeys the following pleasant symmetric laws:

J_0: (a) $\overline{X}$ distinct from X.
(b) $\overline{\overline{X}} = X$.
J_1: (a) The conjugate of any α is some β.
(b) The conjugate of any β is some α.
J_2: (a) $(\overline{\alpha})_1 = \overline{\alpha}_1$; $(\overline{\alpha})_2 = \overline{\alpha}_2$.
(b) $(\overline{\beta})_1 = \overline{\beta}_1$; $(\overline{\beta})_2 = \overline{\beta}_2$.

Let S be a set of unsigned formulas. We leave it to the reader to verify that S is a truth set if and only if S has the following three properties (for every X, α, β): (0) Exactly one of $X, \sim X$ belongs to S. (A) α belongs to S if and only if α_1, α_2 both belong to S. (B) β belongs to S if and only if at least one of β_1, β_2 belong to S.

We shall refer to a set S of signed formulas as a valuation set or truth set if it obeys conditions (A), (B) above and in place of (0), the condition "exactly one of TX, FX belongs to S". We shall also refer to valuation sets of signed formulas as *saturated sets*.

Definition 4.9 An analytic tableaux for X is an ordered dyadic tree, whose points are (occurrences of) formulas, which is constructed as follows. We start by placing X at the origin. Now suppose $\mathcal{T}$ is a tableaux for X which has already been constructed; let Y be an end point. Then we may extend $\mathcal{T}$ by either of the following two operations.

(A) If some P occurs on the path P_y, then we may adjoin either α_1 or α_2 as the sole successor of Y. (In practice, we usually successively adjoin α_1 and then α_2.)

(B) If some P occurs on the path P_y, then we may simultaneously adjoin β_1 as the left successor of Y and β_2 as the right successor of Y.

The above inductive definition of tableaux for X can be made explicit as follows. Given two ordered dyadic trees $\mathcal{T}_1$ and $\mathcal{T}_2$, whose points are occurrences of formulas, we call $\mathcal{T}_2$ a direct extension of $\mathcal{T}_1$ if $\mathcal{T}_2$ can be obtained from $\mathcal{T}_1$ by one application of the operation (A) or (B) above.

Then $\mathcal{T}$ is a tableaux for X iff there exists a finite sequence $(\mathcal{T}_1, \mathcal{T}_2, \ldots, \mathcal{T}_n = \mathcal{T})$ such that $\mathcal{T}_1$ is a 1-point tree whose origin is X and such that for each $i < n$, $\mathcal{T}_{i+1}$ is a direct extension of $\mathcal{T}_i$.

A branch θ of a tableaux for signed (unsigned) formulas is *closed* if it contains some signed formula and its *conjugate* (or some unsigned formula and its negation, if we are working with unsigned formulas.) And $\mathcal{T}$ is called closed if every branch of $\mathcal{T}$ is closed.

By a proof of X is meant a closed tableaux for FX (or for $\sim X$, if we work with unsigned formulas.) It is intuitively rather obvious that any formula provable by the tableaux method must be a tautology–equivalently, given any closed tableau, the origin must be unsatisfiable.

We have thus shown that any immediate extension of a tableaux which is true (under a given interpretation) is again true (under the given interpretation). From this it follows by mathematical induction that for any tableaux $\mathscr{T}$, if the origin is true under a given interpretation v_o, then $\mathscr{T}$ must be true under v_0.

Now a closed tableaux $\mathscr{T}$ obviously cannot be true under any interpretation, hence the origin of a closed tableaux cannot be true under any interpretation–i. e. the origin of any closed tableaux must be unsatisfiable. From this it follows that every formula provable by the tableaux method must be a tautology.

It therefore further follows that the tableaux method is consistent in the sense that no formula and its negation are both provable (since no formula and its negation can both be tautologies).

We now consider the more delicate converse situation: Every tautology is provable by the method of tableaux. Stated otherwise, if X is a tautology, can we be sure that there exists at least one closed tableaux starting with FX?

We might indeed ask the following bolder question: If X is a tautology, then will every complete tableaux for FX close? An affirmative answer to the second question would, of course, be even better than an affirmative answer to the first, since it would mean that any single completed tableaux $\mathscr{T}$ for FX would decide whether X is a tautology or not.

We shall give the proof for tableaux using signed formulas. We are calling a branch of a tableaux complete if for every α which occurs in θ, both α_1 and α_2 occur in θ, and for every β which occurs in θ, at least one of β_1, β_2 occurs in θ.

We call a tableaux $\mathscr{T}$ completed if every branch of $\mathscr{T}$ is either closed or complete. We wish to show that if $\mathscr{T}$ is any completed open tableaux (open in the sense that at least one branch is open), then the origin of $\mathscr{T}$ is satisfiable.

Theorem 4.1 *Any complete open branch of any tableaux is (simultaneously) satisfiable.*

Suppose θ is a complete open branch of a tableaux $\mathscr{T}$; let S be the set of terms of θ. Then the set S satisfies the following three conditions (for every α, β): H_0: No signed variable and its conjugate are both in S). H_1: If $\alpha \in S$, then $\alpha_1 \in S$ and $\alpha_2 \in S$. H_2: If $\beta \in S$, then $\beta_1 \in S$ or $\beta_2 \in S$.

Sets S–whether finite or infinite–obeying conditions H_0, H_1, H_2 are of fundamental importance–we shall call them *Hintikka sets*. We shall also refer to Hintikka sets as sets which are saturated downwards. We shall also call any finite or denumerable sequence θ a *Hintikka sequence* if its set of terms is a Hintikka set.

Lemma 4.1 (Hintikka's Lemma) *Every downward saturated set S (whether finite or infinite) is satisfiable.*

We remark that Hintikka's lemma is equivalent to the statement that every Hintikka set can be extended to a (i. e. is a subset of some) saturated set. We remark that Hintikka's lemma also holds for sets of unsigned formulas (where by a Hintikka set of unsigned formulas we mean a set S satisfying H_1, H_2 and in place of H_0, the condition that no variable and its negation are both elements of S).

Proof Let S be a Hintikka set. We wish to find an interpretation in which every element of S is true. Well, we assign to each variable p, which occurs in at least one element of S, a truth value as follows:

(1) If $Tp \in S$, give p the value true. (2) If $Fp \in S$, give p the value false. (3) If neither Tp nor Fp is an element of S, then give p the value true or false at will (for definiteness, let us suppose we give it the value true.)

We note that the directions (1), (2) are compatible, since no Tp and Fp both occur in S (by hypothesis H_0). We now show that every element of S is true under this interpretation by induction on the degree of the elements.

It is immediate that every signed variable which is an element of S is true under this interpretation (the interpretation was constructed to insure just this). Now consider an element X of S of degree greater than 0, and suppose all elements of S of lower degree than X are true. We wish to show that X must be true. Well, since X is of degree greater than zero, it must be either some α or some β.

Case 1. Suppose it is an α. Then α_1, α_2 must also be in S (by H_1). But α_1, α_2 are of lower degree than α. Hence by inductive hypothesis α_1 and α_2 are both true. This implies that α must be true.

Case 2. Suppose X is some β. Then at least one of β_1, β_2 is in S (by H_2). Whichever one is in S, being of lower degree than β, must be true (by inductive hypothesis). Hence β must be true. This concludes the proof.

Theorem 4.2 (Completeness Theorem for Tableaux) *(a) If X is a tautology, then every completed tableaux starting with FX must close. (b) Every tautology is provable by the tableaux method. To derive statement (a) from Theorem 4.1, suppose $\mathcal{T}$ is a complete tableaux starting with FX. If $\mathcal{T}$ is open, then FX is satisfiable (by Theorem 4.1), hence X cannot be a tautology. Hence if X is a tautology then $\mathcal{T}$ must be closed.*

Let us note that for S, a finite Hintikka set, the proof of Hintikka's lemma 4.1 effectively gives us an interpretation which satisfies S. Therefore, if X is not a tautology, then a completed tableaux for FX provides us with a counterexample of X, in which X is false.

4.3 Tableaux Calculi for Many-Valued Logics

4.3.1 Relationship with Four-Valued Semantics

Our proposed approach is that replacing the base bivalent logic of decision logic with alternative versions of decision logic based on a four-valued semantics. To extend the bivalent semantics of decision logic to a four-valued semantics, we assume a concept of partial semantics for a consequence relation.

Partial semantics for classical logic has been studied by van Benthem in the context of the *semantic tableaux* [25, 27]. Applications of partial semantics for a decision logic of rough sets are investigated in Nakayama et al. [18].

Let $\mathbf{4} = \{\mathbf{T}, \mathbf{F}, \mathbf{N}, \mathbf{B}\}$ be the set of truth values for the four-valued semantics of DL, where each value is interpreted as true, false, neither true nor false, and both true and false, respectively.

A model $\mathcal{M}$ determines a four-valued assignment v on an atomic formula in the following way:

$$\|\varphi\|^v = \begin{Bmatrix} \mathbf{T} \\ \mathbf{F} \\ \mathbf{N} \\ \mathbf{B} \end{Bmatrix} \; \textit{if} \; |\varphi, \sim \varphi|_S \cap S = \begin{Bmatrix} \{\varphi\} \\ \{\sim \varphi\} \\ \{\emptyset\} \\ \{\varphi, \sim \varphi\} \end{Bmatrix}.$$

Then, the truth value of φ on an approximation space $S = (U, R)$ is defined as follows:

$$\|\varphi\|^v = \begin{cases} \mathbf{T} \; \textit{if} \; |\varphi|_S \subseteq POS_R(U/X) \\ \mathbf{F} \; \textit{if} \; |\varphi|_S \subseteq NEG_R(U/X) \\ \mathbf{N} \; \textit{if} \; |\varphi|_S \subseteq BNR_\beta(U/X) \\ \mathbf{B} \; \textit{if} \; |\varphi|_S \subseteq POSR_\beta(U/X) \cap NEGR_\beta(U/X) \end{cases}.$$

where β is precision $\in [0, 0.5)$.

To handle an aspect of partiality on the decision logic, forcing relations for the partial interpretation are defined for four-valued semantics. The truth values of φ are represented by the forcing relation as follows:

$$\begin{aligned} \|\varphi\|^v &= \mathbf{T} \text{ iff } \mathcal{M} \models_v^+ \varphi \text{ and } \mathcal{M} \not\models_v^- \varphi, \\ \|\varphi\|^v &= \mathbf{F} \text{ iff } \mathcal{M} \not\models_v^+ \varphi \text{ and } \mathcal{M} \models_v^- \varphi, \\ \|\varphi\|^v &= \mathbf{N} \text{ iff } \mathcal{M} \not\models_v^+ \varphi \text{ and } \mathcal{M} \not\models_v^- \varphi, \\ \|\varphi\|^v &= \mathbf{B} \text{ iff } \mathcal{M} \models_v^+ \varphi \text{ and } \mathcal{M} \models_v^- \varphi. \end{aligned}$$

A semantic relation for the model $\mathcal{M}$ is defined following Van Benthem [27], Degauquier [9] and Muskens [17].

The truth (denoted by $\models_v^+$) and the falsehood (denoted by $\models_v^-$) of the formulas of the language DL of the decision logic in $\mathcal{M}$ are defined inductively.

Definition 4.10 The semantic relations of $\mathcal{M} \models_v^+ \varphi$ and $\mathcal{M} \models_v^- \varphi$ are defined as follows:

$\mathcal{M} \models_v^+ \varphi$ *iff* $\varphi \in M^+$,

$\mathcal{M} \models_v^- \varphi$ *iff* $\varphi \in M^-$,

$\mathcal{M} \models_v^+ \sim \varphi$ *iff* $\mathcal{M} \models_v^- \varphi$,

$\mathcal{M} \models_v^- \sim \varphi$ *iff* $\mathcal{M} \models_v^+ \varphi$,

$\mathcal{M} \models_v^+ \varphi \vee \psi$ *iff* $\mathcal{M} \models_v^+ \varphi$ *or* $\mathcal{M} \models_v^+ \psi$,

$\mathcal{M} \models_v^- \varphi \vee \psi$ *iff* $\mathcal{M} \models_v^- \varphi$ *and* $\mathcal{M} \models_v^- \psi$,

$\mathcal{M} \models_v^+ \varphi \wedge \psi$ *iff* $\mathcal{M} \models_v^+ \varphi$ *and* $\mathcal{M} \models_v^+ \psi$,

$\mathcal{M} \models_v^- \varphi \wedge \psi$ *iff* $\mathcal{M} \models_v^- \varphi$ *or* $\mathcal{M} \models_v^- \psi$,

$\mathcal{M} \models_v^+ \varphi \to \psi$ *iff* $\mathcal{M} \models_v^- \varphi$ *or* $\mathcal{M} \models_v^+ \psi$,
$\mathcal{M} \models_v^- \varphi \to \psi$ *iff* $\mathcal{M} \models_v^+ \varphi$ *and* $\mathcal{M} \models_v^- \psi$.

The symbol $\sim$ denotes strong negation, in which it is interpreted as true if the proposition is false. Since validity in **B4** is defined in terms of truth preservation, the set of designated values is $\{\mathbf{T}, \mathbf{B}\}$ of **4**.

Example 4.1 The interpretation of formulas in semantics with **B4** is as follows:

$U = \{x_1, x_2, \ldots, x_{20}\}$

Attribute: $A = \{a_1, a_2, a_3, a_4, a_5, a_6\}$, where $a_1 = \{x_1, x_2, x_3, x_4, x_5\}$, $a_2 = \{x_6, x_7, x_8\}$, $a_3 = \{x_9, x_{10}, x_{11}, x_{12}\}$, $a_4 = \{x_{13}, x_{14}\}$, $a_5 = \{x_{15}, x_{16}, x_{17}, x_{18}\}$, $a_6 = \{x_{19}, x_{20}\}$.

$U/A = a_1 \cup a_2 \cup a_3 \cup a_4 \cup a_5 \cup a_6$

Any subset $X = \{x_4, x_5, x_8, x_{14}, x_{16}, x_{17}, x_{18}, x_{19}, x_{20}\}$

$POS_{A^0}(X) = a_6$, where $\beta = 0$
$POS_{A^{0.5}}(X) = a_4 \cup a_5 \cup a_6$, where $\beta = 0.5$
$BN_{A^{0.5}}(X) = a_2 \cup a_4 \cup a_5$, where $\beta = 0.5$
$NEG_{A^0}(X) = a_3$, where $\beta = 0$
$NEG_{A^{0.5}}(X) = a_4$, where $\beta = 0.5$

Evaluation of truth value of formulas as follows:

If $|A_{a_4}|_S \subseteq POS_{A^0}(X)$ *then* $\|A_{a_4}\|^v = \mathbf{T}$,
If $|A_{a_3}|_S \subseteq NEG_{A^0}(X)$ *then* $\|A_{a_3}\|^v = \mathbf{F}$,
If $|A_{a_2}|_S \subseteq BN_{A^{0.5}}(X)$ *then* $\|A_{a_2}\|^v = \mathbf{N}$,
If $|A_{a_4}|_S \subseteq POS_{A^{0.5}}(X) \cap NEG_{A^{0.5}}(X)$ *then* $\|A_{a_4}\|^v = \mathbf{B}$.

4.3.2 Many-Valued Tableaux Calculi

Semantic tableaux can be regarded as a variant of Gentzen systems; see Smullyan [25]. The tableaux calculus is used as the proof method for both classical and non-classical logics in Akama [1] and Priest [24].

The main advantage of the use of the tableaux calculus is that proofs in tableaux calculi are easy to understand. In addition, it is possible to provide a comprehensive argument of completeness proof.

To accommodate the Gentzen system to partial logics, we need some concepts of partial semantics. In the Beth tableaux, It is assumed that V is a partial valuation function assigning to an atomic formula p the values 0 or 1.

We can then set $V(p) = 1$ for p on the left-hand side and $V(p) = 0$ for p on the right-hand side in an open branch of tableaux.

First, we obtain the following concept of consequence relation (4.1) for a classical logic.

$$\text{For all } V, \text{ if } V(Pre) = 1 \text{ then } V(Cons) = 1 \tag{4.1}$$

Pre and *Cons* represent sequent premise and conclusion, respectively and 1 represents true and 0 false.

We use the notion of signed formula. If φ is a formula, then $T\varphi$ and $F\varphi$ are signed formulas. $T\varphi$ reads φ is *provable* and $F\varphi$ reads φ is *not provable*, respectively. If S is a set of signed formulas and α is a signed formula, then we simply define $\{S, \alpha\}$ for $S \cup \{\alpha\}$. As usual, a tableaux calculus consists of axioms and reduction rules. Let p be an atomic formula and φ and ψ be formulas.

The tableaux rules TCL for (4.1) of a propositional classical logic are:

Axiom:

(ID) $\{ Tp, Fp \}$

Tableaux rule:

$$\frac{S, T(\sim \varphi)}{S, F\varphi}\ (T\sim) \quad \frac{S, F(\sim \varphi)}{S, T\varphi}\ (F\sim) \quad \frac{S, T(\varphi \wedge \psi)}{S, T\varphi, T\psi}\ (T\wedge) \quad \frac{S, F(\varphi \wedge \psi)}{S, F\varphi;\ S, F\psi}\ (F\wedge)$$

$$\frac{S, T(\varphi \vee \psi)}{S, T\varphi;\ S, T\psi}\ (T\vee) \quad \frac{S, F(\varphi \vee \psi)}{S, F\varphi, F\psi}\ (F\vee) \quad \frac{S, T(\varphi \rightarrow \psi)}{S, T\varphi;\ S, T\psi}\ (T\rightarrow) \quad \frac{S, F(\varphi \rightarrow \psi)}{S, F\varphi, F\psi}\ (F\rightarrow)$$

A proof of a formula φ is shown with a closed tableaux for $F\varphi$. A tableaux is a tree constructed by the above reduction rules. A tableaux is closed if each branch is closed. A branch is closed if it contains the axioms of the form (ID) in the classical logic. We write $\vdash_{TCL} \varphi$ to mean that φ is provable in TCL.

Theorem 4.3 *The logic for the consequence relation (4.1) is axiomatized by the tableaux calculus TCL.*

Proof See Smullyan [25], Van Benthem [27], and Akama [1].

The tableaux calculus TCL* is extended TCL excluded rules of $(T \sim)$ and $(F \sim)$ by introducing axioms of the principle of explosion EFQ (ex falso quodlibet) and the excluded middle LEM, and the following rules:

(EFQ) $\{Tp, T \sim p\}$

(LEM) $\{Fp, F \sim p\}$

$$\frac{S, T(\sim (\varphi \wedge \psi))}{S, T(\sim \varphi);\ S, T(\sim \psi)}\ (T \sim \wedge) \qquad \frac{S, F(\sim (\varphi \wedge \psi))}{S, F(\sim \varphi), F(\sim \psi)}\ (F \sim \wedge)$$

$$\frac{S, T(\sim (\varphi \vee \psi))}{S, T(\sim \varphi), T(\sim \psi)}\ (T \sim \vee) \qquad \frac{S, F(\sim (\varphi \vee \psi))}{S, F(\sim \varphi);\ S, F(\sim \psi)}\ (F \sim \vee)$$

$$\frac{S, T(\sim (\varphi \rightarrow \psi))}{S, T\varphi, T(\sim \psi)}\ (T \sim\rightarrow) \qquad \frac{S, F(\sim (\varphi \rightarrow \psi))}{S, F\varphi;\ S, F(\sim \psi)}\ (F \sim\rightarrow)$$

$$\frac{S, T(\sim\sim \varphi)}{S, T\varphi}\ (T \sim\sim) \qquad \frac{S, F(\sim\sim \varphi)}{S, F\varphi}\ (F \sim\sim)$$

Next, we extend consequence relation (4.2) for Belnap's logic **B4** as follows:

$$\textit{For all } V,\ \textit{if } V(Pre) \neq 0 \textit{ then } V(Cons) \neq 0. \tag{4.2}$$

Equations (4.1) and (4.2) are not different as the formulation of classical validity. However, they should be distinguished for partial semantics with four-valued interpretation.

The tableaux calculus TC4 for (4.2) is defined from TCL* without $(T \sim)$, $(F \sim)$, (EFQ) and (LEM) as follows:

TC4 $:= \{(ID),\ (T\wedge),\ (F\wedge),\ (T\vee),\ (F\vee),\ (T \rightarrow),\ (F \rightarrow),\ (T \sim\wedge),\ (F \sim\wedge),$
$(T \sim\vee),\ (F \sim\vee),\ (T \sim\rightarrow),\ (F \sim\rightarrow),\ (T \sim\sim),\ (F \sim\sim)\}.$

Equation (4.2) is regarded as a four-valued logic since it allows for incomplete and inconsistent valuation. We define the extension of the valuation function $v(p)$ for an atomic formula p as follows:

Definition 4.11 (Valuation function v for C)
$\mathbf{T} =_{\text{def}} v(p) = 1 =_{\text{def}} v(p) = 1 \textit{ and } v(p) \neq 0,$
$\mathbf{F} =_{\text{def}} v(p) = 0 =_{\text{def}} v(p) = 0 \textit{ and } v(p) \neq 1,$
$\mathbf{N} =_{\text{def}} v(p) = \{\} =_{\text{def}} v(p) \neq 1 \textit{ and } v(p) \neq 0,$
$\mathbf{B} =_{\text{def}} v(p) = \{1, 0\} =_{\text{def}} v(p) = 1 \textit{ and } v(p) = 0.$

In this valuation, the law of contradiction fails since if we have $p\wedge \not p$ in premises, both p and $\not p$ are evaluated as **B**. Additionally, the law of excluded middle fails since if we have $p\vee \not p$ in conclusions, both p and $\not p$ are evaluated as **N**.

Now, we try to extend TC4 with weak negation and weak implication. Weak implication regains the deduction theorem that some many-valued logics lack.

Here, we introduce weak negation "¬". The semantic relation for weak negation is as follows:

$$\mathcal{M} \models_v^+ \not\varphi \text{ iff } \mathcal{M} \not\models_v^+ \varphi,$$
$$\mathcal{M} \models_v^- \not\varphi \text{ iff } \mathcal{M} \models_v^+ \varphi.$$

A semantic interpretation of weak negation is denoted as follows:

$$\|\neg\varphi\|^v = \begin{cases} \mathbf{F} \text{ if } \|\varphi\|^v = \mathbf{T} \text{ or } \mathbf{B} \\ \mathbf{T} \text{ if } \|\varphi\|^v = \mathbf{F} \text{ or } \mathbf{N} \end{cases}.$$

Weak negation can represent the absence of truth and the reading for $\not\varphi$ is as "$\not\varphi$ is not *true*". However, "$\sim$" can serve as strong negation to express the verification of falsity.

Next, weak implication is defined as follows:

$$\varphi \rightarrow_w \psi =_{\text{def}} \neg\varphi \vee \psi.$$

A semantic interpretation of weak implication is defined as follows:

$$\|\varphi \rightarrow_w \psi\|^v = \begin{cases} \|\psi\|^v \text{ if } \|\varphi\|^v \in \mathcal{D} \\ \mathbf{T} \textit{ if } \|\varphi\|^v \notin \mathcal{D} \end{cases}.$$

Unlike "$\rightarrow$", weak implication satisfies the deduction theorem. This means that it can be regarded a logical implication. We can also interpret weak negation in terms of classical negation and weak implication:

$$\neg\varphi =_{\text{def}} \varphi \rightarrow_w \sim \varphi.$$

We extend TC4 with weak negation and weak implication as TC4$^+$. So, we define the tableaux rules for $(\neg)$ and $(\rightarrow_w)$ as follows:

$$\frac{S, T(\neg\varphi)}{S, F\varphi}\ (T\neg) \quad \frac{S, F(\neg\varphi)}{S, T\varphi}\ (F\neg) \quad \frac{S, T(\varphi \rightarrow_w \psi)}{S, F\varphi;\ S, T\psi}\ (T{\rightarrow_w}) \quad \frac{S, F(\varphi \rightarrow_w \psi)}{S, T\varphi, F\psi}\ (F{\rightarrow_w})$$

Here, the tableaux calculus TC4$^+$ is defined as follows:

TC4$^+$ $:= \{(ID),\ (T\wedge),\ (F\wedge),\ (T\vee),\ (F\vee),\ (T \rightarrow),\ (F \rightarrow),\ (T \sim\wedge),$
$(F \sim\wedge), (T \sim\vee),\ (F \sim\vee), (T \sim\rightarrow),\ (F \sim\rightarrow),\ (T \sim\sim),\ (F \sim\sim),\ (T\neg),$
$(F\neg),\ (T \rightarrow_w T),\ (F \rightarrow_w)\}.$

TC4$^+$ is interpreted as an extended four-valued logic with weak negation and weak implication.

Next, we discuss another extension of TC4$^+$. We extend them by adding a unary operator presented in Epstein [11], which is usually used to represent *crispness* in the sense of rough sets. Here, we only describe the definition and semantic interpretation.

$$©\varphi =_{\text{def}} \neg\neg\varphi \vee \neg\neg \sim \varphi.$$

$$\|\copyright\varphi\|^v = \begin{cases} \mathbf{T} \text{ if } \|\varphi\|^v = \mathbf{T},\ \mathbf{F}\ or\ \mathbf{B} \\ \mathbf{F} \text{ if } \|\varphi\|^v = \mathbf{N} \end{cases}.$$

This operator is useful to handle undefined information in decision logic.

4.4 Soundness and Completeness

In this section, the soundness and completeness are proved for the tableaux system TC4$^+$. A *proof* of a formula φ is a closed tableaux for $F\varphi$. A tableaux is a tree constructed by the reduction rules defined in previous section. A tableaux is closed if each branch is closed, where it contains the axiom of the form (ID).

We write $\vdash_{TC4^+} \varphi$ to mean that φ is provable in TC4$^+$. We see that φ is true iff $v(\varphi) = 1$. φ is valid, written $\models_{TC4^+} \varphi$, iff it is true in all four-valued models of Belnap's logic **B4**. We prove the completeness of the tableaux calculus TC4$^+$ with respect to Belnap's four-valued semantics. The proof strategy is similar to the way sketched in Akama [1].

Let $S = \{T\varphi_1, \ldots, T\varphi_n, F\psi_1, \ldots, F\psi_m\}$ be a set of signed formula, $\mathcal{M}$ be a four-valued with weak negation model. We say that valuation v refutes S if

$$v(\varphi_i) = 1 \text{ if } T\varphi_i \in S,$$
$$v(\psi_i) \neq 1 \text{ if } F\psi_i \in S.$$

A set S is *refutable* if something refutes it. If S is not refutable, it is *valid.*

Theorem 4.4 (Soundness of TC4$^+$) *If φ is provable, then φ is valid.*

Proof For any formula φ in TC4$^+$, the following holds: $\vdash_{TC4^+} \varphi$ iff $\models_{TC4^+} \varphi$. If φ is of the form of axioms, it is easy to see that it is valid. For reduction rules, it suffices to check that they preserve validity. We only show the cases of $(T \sim \vee)$ and $(F \sim\rightarrow)$.

$(T \sim \vee)$: We have to show that if $S, T(\sim (\varphi \vee \psi))$ is refutable then $S, T(\sim \varphi), T(\sim \psi)$ is also refutable. By the assumption, there is a semantic relation Definition 4.10, in which valuation v refutes S and $\models^+_v \sim (\varphi \vee \psi)$. This implies:

$$\begin{aligned} v(\varphi \vee \psi) \neq 1 \text{ iff } & v(\varphi) \neq 1 \text{ and } v(\psi) \neq 1 \\ \text{iff } & v(\varphi) = 0 \text{ and } v(\psi) = 0 \\ \text{iff } & v(\sim \varphi) = 1 \text{ and } v(\sim \psi) = 1. \end{aligned}$$

Therefore, $S, T(\sim \varphi), T(\sim \psi)$ is shown to be refutable.

($F \sim\to$): By the assumption, there is a semantic relation of the weak negation, which refutes S and $\models_v^- \sim (\varphi \to \psi)$. This implies:

$$v(\varphi \to \psi) \neq 0 \text{ iff } v(\varphi) \neq 1 \text{ and } v(\sim \psi) \neq 1.$$

Therefore, $S, F\varphi$ and $S, F(\sim \psi)$ are refutable. We can show other cases.

We are now in a position to prove completeness of TC4$^+$. The proof below is similar to the Henkin proof described in Akama [1], which is extended for paraconsistent logic.

A finite set of signed formulas Γ is *non-trivial* if no tableaux for it is closed. An infinite set of signed formulas is non-trivial if every finite subset is non-trivial. If a set of formulas is not non-trivial, it is trivial. Every formula is provable from a trivial set.

Lemma 4.2 *A non-trivial set of signed formulas Γ_0 can be extended to a maximally non-trivial set of signed formulas Γ.*

Proof Since the language L has a countably infinite set of sentences, we can enumerate sentences $\varphi_1, \varphi_2, \dots \varphi_n$. Now, we define for a non-trivial set of signed formulas Γ_0 a sequence of non-trivial sets of signed formulas $\Gamma_0, \Gamma_1, \dots \Gamma_n$ in the following way:

$$\Gamma_{n+1} = \begin{cases} \Gamma_n \cup \{T\varphi_{n+1}\} \text{ if } \Gamma_n \cup \{T\varphi_{n+1}\} \text{ is non-trivial,} \\ \Gamma_n \cup \{F\varphi_{n+1}\} \text{ if } \Gamma_n \cup \{F\varphi_{n+1}\} \text{ is non-trivial,} \\ \Gamma_n \qquad\qquad\qquad otherwise. \end{cases}$$

Then we obtain:

$$\Gamma = \bigcup \Gamma_i$$

It is obvious that Γ satisfies the desired properties of a maximally non-trivial set.

We here define a *canonical model* with respect to the tableaux TC4$^+$:

Definition 4.12 Based on the maximal non-trivial set, we can define a canonical model $(\Gamma, \subseteq, V)$ such that Γ is a tableaux TC4$^+$, $\subseteq$ is the subset relation, and V is a valuation function satisfying the condition that $V(\varphi, \Gamma) = 1$ iff $\varphi \in \Gamma$ and that $V(\varphi, \Gamma) = 0$ iff $\sim \varphi \in \Gamma$.

A canonical model for TC4$^+$ satisfies the desired properties of the consequence relation for four-valued semantics.

Lemma 4.3 *For any $\Gamma \in S$ in a canonical model $(\Gamma, \subseteq, V)$ we have the following properties:*

(1) if $T(\varphi \wedge \psi) \in \Gamma$, then $T\varphi \in \Gamma$ and $T\psi \in \Gamma$,
(2) if $F(\varphi \wedge \psi) \in \Gamma$, then $F\varphi \in \Gamma$ or $F\psi \in \Gamma$,
(3) if $F(\varphi \vee \psi) \in \Gamma$, then $T\varphi \in \Gamma$ or $T\psi \in \Gamma$,
(4) if $T(\varphi \vee \psi) \in \Gamma$, then $F\varphi \in \Gamma$ and $F\psi \in \Gamma$,
(5) if $T(\varphi \rightarrow \psi) \in \Gamma$, then $F\varphi \in \Gamma$ or $T\psi \in \Gamma$,
(6) if $F(\varphi \rightarrow \psi) \in \Gamma$, then $T\varphi \in \Gamma$ and $F\psi \in \Gamma$,
(7) if $T(\sim (\varphi \wedge \psi)) \in \Gamma$, then $T(\sim \varphi) \in \Gamma$ or $T(\sim \psi) \in \Gamma$,
(8) if $F(\sim (\varphi \wedge \psi)) \in \Gamma$, then $F(\sim \varphi) \in \Gamma$ and $F(\sim \psi) \in \Gamma$,
(9) if $T(\sim (\varphi \vee \psi)) \in \Gamma$, then $T(\sim \varphi) \in \Gamma$ and $T(\sim \psi) \in \Gamma$,
(10) if $F(\sim (\varphi \vee \psi)) \in \Gamma$, then $F(\sim \varphi) \in \Gamma$ or $F(\sim \psi) \in \Gamma$,
(11) if $T(\sim\sim \varphi) \in \Gamma$, then $T\varphi \in \Gamma$,
(12) if $F(\sim\sim \varphi) \in \Gamma$, then $F\varphi \in \Gamma$,
(13) if $T(\neg\varphi) \in \Gamma$, then $F\varphi \in \Gamma$,
(14) if $F(\neg\varphi) \in \Gamma$, then $T\varphi \in \Gamma$,
(15) if $T(\varphi \rightarrow_w \psi) \in \Gamma$, then $T\varphi \notin \Gamma$ or $T\psi \in \Gamma$,
(16) if $F(\varphi \rightarrow_w \psi) \in \Gamma$, then $F\varphi \notin \Gamma$ and $F\psi \in \Gamma$.

Proof See Akama [1].

Theorem 4.5 *For any $\Gamma \in S$ in a canonical model and any formula φ,*

$T\varphi \in \Gamma$ iff $V(\Gamma, \varphi) \neq 0$,
$F\varphi \in \Gamma$ iff $V(\Gamma, \varphi) \neq 1$.

Proof (We only show the cases of $\wedge$, $\rightarrow$ and $\rightarrow_w$.) By induction on a formula Σ. The case Σ is an atomic formula is immediate.
(1) $\Sigma = \varphi \wedge \psi$ (also $\sim (\varphi \vee \psi)$):

$$\begin{aligned} T(\varphi \wedge \psi) \in \Gamma \text{ iff } & T\varphi \in \Gamma \text{ and } T\psi \in \Gamma \\ \text{iff } & V(\Gamma, \varphi) \neq 0 \text{ and } V(\Gamma, \psi) \neq 0 \\ \text{iff } & V(\Gamma, \varphi \wedge \psi) \neq 0. \\ F(\varphi \wedge \psi) \in \Gamma \text{ iff } & F\varphi \in \Gamma \text{ or } F\psi \in \Gamma \\ \text{iff } & V(\Gamma, \varphi) \neq 1 \text{ or } V(\Gamma, \psi) \neq 1 \\ \text{iff } & V(\Gamma, \varphi \wedge \psi) \neq 1. \end{aligned}$$

(2) $\Sigma = \varphi \rightarrow \psi$:

$$
\begin{aligned}
T(\varphi \to \psi) \in \Gamma \text{ iff } & (F\varphi \in \Gamma \text{ or } T\psi \in \Gamma) \text{ or} \\
& ((T\varphi \in \Gamma \text{ and } F\varphi \in \Gamma) \text{ and} \\
& (T\psi \notin \Gamma \text{ and } F\psi \notin \Gamma)) \text{ or} \\
& ((T\varphi \notin \Gamma \text{ and } F\varphi \notin \Gamma) \text{ and} \\
& (T\psi \in \Gamma \text{ and } F\psi \in \Gamma))
\end{aligned}
$$

$$
\begin{aligned}
\text{iff } & (V(\Gamma, \varphi) = 0 \text{ or } V(\Gamma, \psi) = 1) \text{ or} \\
& ((V(\Gamma, \varphi) = 1 \text{ and } V(\Gamma, \varphi) = 0) \text{ and} \\
& (V(\Gamma, \psi) \neq 1 \text{ and } V(\Gamma, \psi) \neq 0)) \text{ or} \\
& (V(\Gamma, \varphi) \neq 1 \text{ and } V(\Gamma, \varphi) \neq 0) \text{ and} \\
& (V(\Gamma, \psi) = 1 \text{ and } V(\Gamma, \psi) = 0)) \\
\text{iff } & (V(\Gamma, \varphi \to \psi) = 1.
\end{aligned}
$$

$$
\begin{aligned}
F(\varphi \to \psi) \in \Gamma \text{ iff } & T\varphi \in \Gamma \text{ and } F\psi \in \Gamma \\
\text{iff } & V(\Gamma, \varphi) = 1 \text{ or } V(\Gamma, \psi) = 0 \\
\text{iff } & V(\Gamma, \varphi \to \psi) = 0.
\end{aligned}
$$

(3) $\Sigma = \varphi \to_w \psi$:

$$
\begin{aligned}
T(\varphi \to_w \psi) \in \Gamma \text{ iff } & T\varphi \notin \Gamma \text{ or } T\psi \in \Gamma \\
\text{iff } & V(\Gamma, \varphi) \neq 1 \text{ or } V(\Gamma, \psi) = 1 \\
\text{iff } & V(\Gamma, \varphi \to_w \psi) = 1. \\
F(\varphi \to_w \psi) \in \Gamma \text{ iff } & F\varphi \notin \Gamma \text{ and } F\psi \in \Gamma \\
\text{iff } & V(\Gamma, \varphi) \neq 0 \text{ and } V(\Gamma, \psi) = 0 \\
\text{iff } & V(\Gamma, \varphi \to_w \psi) = 0.
\end{aligned}
$$

As a consequence, we show the completeness of TC4$^+$:

Theorem 4.6 (Completeness Theorem) $\vdash_{TC4^+} \varphi$ *iff* $\models_{TC4^+} \varphi$.

Proof The soundness was already proved in Theorem 4.2. For the completeness, it suffices to show that an open tableaux is refutable by a counter model by theorem 4.3. By contraposition, the completeness theorem follows.

4.5 Concluding Remarks

In this chapter we have formalized four-valued tableaux calculi for decision logic. As a semantics for inconsistent data table, we introduce Belnap's four-valued logic [6, 7].

Four-valued semantics is applied to extend decision logic for inconsistent data table. We also provided Variable Precision Rough Set model to define four-valued semantics for decision logic. Furthermore, we extend the four-valued tableaux calculi with weak negation to repair deduction of four-valued logic.

There are some topics that can be further developed. First, it is very interesting to apply another kind of proof system or deduction method to decision logic. We are also interested to apply another kind of semantics and models to interpret decision logic that includes inconsistent information.

Second, we need to extend the present work for the predicate logic for decision logic, and also need to show completeness. Third, we need to investigate applications of decision logic with four-valued semantics for inconsistent information.

References

1. Akama, S.: Nelson's paraconsistent logics. Logic Logical Philos. 7(1999), 101–115 (1999)
2. Akama, S., Murai, T., Kudo, Y.: Reasoning with Rough Sets. Springer, Heidelberg (2018)
3. Akama, S., Nakayama, Y.: Consequence relations in DRT. In: Proc. of the 15th International Conference on Computational Linguistics (COLING 1994) 2, pp. 1114–117 (1994)
4. Anderson, A., Belnap, N.: Entailment: The Logic of Relevance and Necessity I. Princeton University Press, Princeton (1976)
5. Avron, A., Konikowska, B.: Rough sets and 3-valued logics. Studia Logica **90**, 69–92 (2008)
6. Belnap, N.D.: A useful four-valued logic. In: Dunn, J.M., Epstein, G. (eds.) Modern Uses of Multi-Valued Logic, pp. 7–37. Reidel, Dordrecht (1977)
7. Belnap, N.D.: How a computer should think. In: Ryle, G. (ed.) Contemporary Aspects of Philosophy, pp. 30–55. Oriel Press (1977)
8. Ciucci, D., Dubois, D.: Three-valued logics, uncertainty management and rough sets. In: Transactions on Rough Sets XVII. Lecture Notes in Computer Science Book Series (LNCS), 8375, pp. 1–32. Springer, Berlin (2001)
9. Degauquier, V.: Partial and paraconsistent three-valued logics. Logic Logical Philos. **25**, 143–171 (2016)
10. Doherty, P.: NM3–A three-valued cumulative non-monotonic formalism. In: van Eijck (ed.) Proc. of European Workshop on Logics in Artificial Intelligence (JELIA), pp 196–211. Springer, Heidelberg (1990)
11. Epstein, R.L.: The Semantic Foundations of Logic. Springer, Heidelberg (1990)
12. Fan, T.-F., Hu, W.-C., Liau, C.-J.: Decision logics for knowledge representation in data mining. In: Proc. of the 25th Annual International Computer Software and Applications Conference (COMPSAC), pp. 626–631 (2001)
13. Gentzen, G.: Untersuchungen über das logische Schliesen. I. Math. Z. **39**, 176–210 (1935)
14. Lin, Y., Qing, L.: A logical method of formalization for granular computing. In: Proc. of the IEEE International Conference on Granular Computing (GRC 2007), pp. 22–27 (2007)
15. Łukasiewicz, J.: On 3-valued logic, 1920. In: McCall, S. (ed.) Polish Logic, pp. 16–18. Oxford University Press, Oxford (1967)

16. Łukasiewicz, J.: Many-valued systems of propositional logic, 1930. In: McCall, S. (ed.) Polish Logic. Oxford University Press, Oxford (1967)
17. Muskens, R.: On partial and paraconsistent logics. Notre Dame J. Formal Logic **40**, 352–374 (1999)
18. Nakayama, Y., Akama, S., Murai, T.: Deduction system for decision logic based on partial semantics. In: Proc. of the 11th International Conference on Advances in Semantic Processing, pp. 8–11 (2017)
19. Nakayama, N., Akama, S., Murai, T.: Four-valued tableau calculi for decision logic of rough set. In: Proc. of KES-2018, pp. 383–392. Elsevier, Amsterdam (2018)
20. Nakayama, Y., Akama, S., Murai, T.: Deduction system for decision logic based on many-valued Logics. Int. J. Adv. Intell. Syst. **11**, 115–126 (2018)
21. Pawlak, Z.: Rough sets. Int. J. Comput. Inf. Sci. **11**, 341–356 (1982)
22. Pawlak, Z.: Rough Sets: Theoretical Aspects of Reasoning about Data. Kluwer, Dordrecht (1991)
23. Priest, G.: The logic of paradox. J. Philos. Logic **8**, 219–241 (1979)
24. Priest, G.: An Introduction to Non-Classical Logic, From If to Is, 2nd edn. Cambridge University Press, Cambridge (2008)
25. Smullyan, R.: First-Order Logic. Springer, Berlin (1968)
26. Urquhart, A.: Basic many-valued logic. In: Gabbay, G., Guenthner, F. (eds.) Handbook of Philosophical Logic, 2, pp. 249–295. Springer, Heidelberg (2001)
27. van Benthem, J.: Partiality and nonmonotonicity in classical logic. Logique et Analyse **29**, 225–247 (1986)
28. Vitoria, A., Andrzej, A.S., Maluszynski, J.: Four-valued extension of rough sets. In: Proc. of the International Conference on Rough Sets and Knowledge Technology (RSKT), pp. 106–114 (2008)
29. Ziarko, W.: Variable precision rough set model. J. Comput. Syst. Sci. **46**, 39–59 (1993)

Chapter 5
Granular Reasoning for the Epistemic Situation Calculus

Abstract In Chap. 5, we apply granular reasoning to the epistemic situation calculus ES of Lakemeyer and Levesque by interpreting actions as modalities and granules of possible worlds as states. The zoom reasoning proposed by Murai et al. is regarded as an epistemic action and is incorporated into the ES as an abstraction and refinement action by the granularity of the situation.

5.1 Introduction

This chapter is an extended version of Nakayama et al. [20]. Murai et al. [16–18] have proposed a framework of granular reasoning called *zooming reasoning* based on granular computing. In zooming reasoning, a filtration method is used to control the degree of granularity based on the rough set.

This dynamic feature of the degree of granularity of zooming reasoning serves the mechanism of dynamic semantic interpretation for abstraction and refinement. Zooming reasoning is possible to apply for non-monotonic reasoning as well.

In *epistemic situation calculus* (ES) by Lakemeyer and Levesque [13, 14], to incorporate the modal logic into the situation calculus [15, 26], it enables to formulate the epistemic state of an agent, and the situation is interpreted as a possible world in ES.

The key concept of the zooming reasoning system is focus, which represents sentences we use in the current step of reasoning.

The focus provides "granularized" possible worlds, and a four-valued valuation gives an interpretation of a situation. The truth value **T** means just told True, **F** means just told False, **N** means told neither True nor False, and **B** means told both True and False. In addition, Murai et al. have provided a mechanism of control of the degree of granularity,

In this chapter, we propose to incorporate the epistemic action of abstraction and refinement to the epistemic situation calculus ES, and these two epistemic actions are represented with zooming reasoning as the action of ES. Therefore, we assume that a epistemic thinking pattern of abstraction and refinement as epistemic action in ES.

S. Akama et al., *Epistemic Situation Calculus Based on Granular Computing*, Intelligent Systems Reference Library 239,
https://doi.org/10.1007/978-3-031-28551-6_5

By capturing the zooming reasoning as actions which change the epistemic state of an agent, from the perspective of possible worlds and granularization of worlds, on the framework of the epistemic situation calculus ES, it is natural to incorporate zooming reasoning, and also it is expected that this enhances the expression power of ES.

As the semantics basis of the zooming reasoning, we employ the interpretation by a four-valued logic of Belnap [5, 6], and the interpretation of modal logic with four-valued semantics [23]. In addition, as the basis of the deductive system, we utilize the axiomatization with the sequent calculi using the decision logic of rough set theory.

Our research relates the following theories, such as epistemic situation calculus, dynamic epistemic logic, granular computing, and non-classical logics such as modal logics and many-valued logics. Besides, we describe related previous researches as follows:

In Nakayama et al. [21], as for the relationship between situation calculus and zooming reasoning based on granular computing, the deduction system with four-valued logic is studied. Demolombe [8] proposed a method using explicit frame axioms to give a transformation from situational calculus to dynamic logic.

Ditmarsch et al. [10] studied the correspondence between situation calculus and *dynamic epistemic logic* corresponding to *public announcement logic*.

As for granular reasoning based on modal logic and rough sets, Kudo et al. [12] propose a granularity-based framework of abduction using variable precision rough set models (VPRS) and measure-based semantics for modal logic. The relationship between conditional logic and granular reasoning is discussed in Murai [19], and also their applications are studied.

In addition, Banihashemi [4] described the high level and low level action theory for the abstraction for the action theory, and also proposed a mapping method for the high level and the low level action theory.

Nakayama et al. [21] studied the application of many-valued logics as a basis of a deduction for non-monotonic reasoning in the situation calculus. Akama et al. [1] surveyed researches in the field of rough sets and granular reasoning.

The structure of this chapter is as follows. In Sect. 5.2, the outline of epistemic situation calculus ES is explained, and in Sect. 5.3, the rough set, its decision logic as a deductive system, and VPRS are explained. Section 5.4 describes the outline of a possible world model and an application of granular worlds to zooming reasoning, and Sect. 5.5 discusses deduction systems based on four-valued logic and sequential calculus. Section 5.6 outlines the application of zooming reasoning as an action of ES, and finally summarizes and discusses future issues.

5.2 Epistemic Situation Calculus

5.2.1 Background of Lakemeyer & Levesque's Logic ES

Here, we describe the outline of the epistemic situation calculus, which incorporates a modal logic framework into the situation calculus. The situation calculus is a system of the first order logic for the temporal reasoning where there are two sorts that are the situation s and the action a and the function do refer the successor situation $do(a, s)$ which is obtained as the result where the action a is performed at the situation s.

The *situation* is the state of the world at a specific time, and the quantity which is changed according to the time is called *fluent* and represented as the function of the situation.

When a fluent is a proposition, its domain of the fluent is the truth value, and if a function, then its domain is various value; for example, if it takes a continuous quantity then the domain is a real number [15]. The situation calculus is based on a discrete and finite transition model.

In the situation calculus, basic elements for the description target is as follow:

- *Situations*: the complete state of the world at an instantaneous time.
- *Fluent*: a function with a set of circumstances composed of situations as a domain, a propositional fluent whose value range is a boolean value, and a situation whose value range is a situation value is called a contextual fluent.
- *Actions*: actions performed worldwide. The combination of actions is called a strategy. For the contextual fluence, a new situation arises as a result of the action performed.

In addition, the situation calculus defines an action precondition axioms, a successor state axioms for each action and each fluent. The action precondition axiom defines the precondition using a specific predicate $Poss(a, s)$, which is required at the execution of an action on some situation.

The parameter a is specified for the action and s the situation. The successor state axiom is a generalization of the frame axiom and defines a possible result of a fluent after the execution of an action.

In these descriptions, it is suitable for inferring the influence of action, but it cannot deduce that it will not be affected. For this reason, a frame axiom is defined to describe that when a specific action does not change a specific fluent, it does not change.

However, if the frame axiom becomes a large scale, it is difficult for the programmer to deal with everything appropriately. As an approach to the frame problem, formalization of the partiality to the dynamic world is essential.

Epistemic situation calculation ES is an extension of situation calculus with epistemic logic [14], and was proposed by Lakemeyer and Levesque [13].

Language ES: In epistemic situation calculus ES, $\mathscr{A}$ is a non-empty action variable, $\mathscr{F}$ is fluent, A is an action constant, predicate *Poss* (*possible*), and *SF* (*sensed fluent*).

The language $\mathscr{L}_{ES}$ is represented by the following BNF.

$$\varphi ::= p \mid Poss(t) \mid SF(t) \mid t = t \mid \sim \varphi \mid \varphi \wedge \psi \mid \varphi \vee \psi \mid \varphi \rightarrow \psi \mid \mathbf{K}\varphi \mid [t]\varphi \mid \Box\varphi \mid \forall x \varphi$$

The predicate *Poss* models executability preconditions of actions. The intuition is that $Poss(a)$ is used to abbreviate a formula which holds if and only if the action a is executable, where p ranges over $\mathscr{F}$, t ranges over $\mathscr{A} \cup A$, and x over $\mathscr{A}$.

The predicate SF is used to model the result of actions. The formula $SF(a)$ abbreviates a formula whose truth-value is known by the agent after the execution of a.

As in epistemic logics, the operator **K** models the agent's knowledge. The operator $[\cdot]$ is used to model actions. A formula of the form $[a]\varphi$ is read that it holds after the execution of a. The formula $\Box\varphi$ is read that it holds after the execution of any sequence of actions.

5.2.2 *Semantics of ES*

The main purpose of the semantics of ES is to represent fluents, which may vary as the result of actions and whose values may be unknown.

Let A^* be the set of all sequences of actions from A, where $[\cdot]$ denotes the empty sequence and $\alpha \cdot \alpha^*$ is the concatenation of α and α^*.

A world $w \in W$ is any function from the primitive sentences and action A to $\{0, 1\}$.[1] Intuitively, to determine whether or not a sentence φ is true after a sequence of actions z has been performed, for a world w and an epistemic state e, a sentence φ without free variable is true is written:

$$e, w, z \models \varphi$$

A world determines truth values for the primitive sentences and the primitive terms after any sequence of actions. An epistemic state is defined by a set of worlds, as in possible world semantics.

To interpret what is known by the agent after a sequence of actions, Lakemeyer and Levesque inductively define an *indistinguishable relation* or *agreement relation* of two worlds with respect to a sequence of actions:

1. $w' \simeq_{()} w$ for all $w', w \in W$
2. $w' \simeq_{z,n} w$ iff $w' \simeq_z w$ and $w'(SF(n), z) = w(SF(n), z)$

[1] In Lakemeyer and Levesque [13], ES contains the definition of functional fluent that we do not consider here.

That is, w and w' are indistinguishable after action a if they were so before, and if n's sensed fluent has the same value at w and w' before n.

The semantic relation $\models$ of ES is defined inductively by:

$\langle e, w, z\rangle \models \varphi$ iff $w(\varphi, z) = 1$ if φ is a ground atomic formula.
$\langle e, w, z\rangle \models t_1 = t_2$ iff t_1 and t_2 are identical.
$\langle e, w, z\rangle \models \sim \varphi$ iff $\langle e, w, z\rangle \not\models \varphi$.
$\langle e, w, z\rangle \models \varphi \wedge \psi$ iff $\langle e, w, z\rangle \models \varphi$ and $\langle e, w, z\rangle \models \psi$.
$\langle e, w, z\rangle \models \mathbf{K}\varphi$ iff for all $w' \in e$, if $w' \simeq_z w$ then $\langle e, w', z\rangle \models \varphi$,
$\langle e, w, z\rangle \models [n]\varphi$ iff $\langle e, w, z \cdot n\rangle \models \varphi$.$\langle e, w, z\rangle \models \Box\varphi$ iff for all $z' \in \mathscr{Z}$, $\langle e, w, z \cdot z'\rangle \models \varphi$.
$\langle e, w, z\rangle \models \forall x\varphi$ iff for all n, $\langle e, w, z\rangle \models \varphi[x\backslash a]$.

A formula $\varphi \in \mathscr{L}_{ES}$ is a valid ES consequence of a set of formulas $\Psi \subseteq \mathscr{L}_{ES}$, noted $\Psi \models_{ES} \varphi$, if and only if for all e and w, if $e, w \models \psi$ for all $\psi \in \Psi$ then $e, w \models \varphi$. A formula φ is ES valid, noted $\models_{ES} \varphi$, if and only if $\emptyset \models_{ES} \varphi$.

Knowledge: All elements of the subset e of given possible worlds need not be true. In addition, e represents the initial state of knowledge and another knowledge is obtained according to the execution of an action, and the part of knowledge may become not true. Therefore the logic of knowledge is assumed to be weak S5 or K45 [14].

5.2.3 Basic Action Theory

As shown in Lakemeyer and Levesque [13], we are able to define basic action theories in a way very similar to those originally introduced by Reiter:

Definition 5.1 *(Basic Action Theory)* Given a set of fluent predicates F, a set of sentences Σ is called a basic action theory over $\mathscr{F}$ iff it only mentions the fluents in $\mathscr{F}$ and is of the form $\Sigma = \Sigma_0 \cup \Sigma_{pre} \cup \Sigma_{post} \cup \Sigma_{sense}$, where

1. Σ_0 is a finite set of fluent sentences;
2. Σ_{pre} contains a single sentence $\Box Poss(a) \equiv \mathscr{F}$ where $\mathscr{F}$ is a fluent formula;
3. Σ_{post} contains a sentence $\Box F(\mathbf{x}) \equiv \gamma_F$ for all $F \in \mathscr{F}$ where γ_F is a fluent formula;
4. Σ_{sense} contains a single sentence $\Box SF(a) \equiv \varphi$ where φ is a fluent formula.

The idea here is that Σ_0 expresses what is true initially (in the initial situation), Σ_{pre} is one large precondition axiom, and Σ_{post} is a set of successor state axioms, one per fluent, and Σ_{sense} defines the sensing results for actions.

We adopt Basic Action Theory according to Schwering and Lakemeyer [27]. We use the modal variant of Reiter's basic action theories to axiomatize a dynamic domain [26].

A basic action theory over a finite set of fluent $\mathscr{F}$ consists of a static and a dynamic part. The concept of dynamic axioms represents an action precondition (Σ_{pre}), changed truth-value of fluents after actions (Σ_{post}), and knowledge sensed after actions (Σ_{sense}):

The successor state axiom of knowledge is represented as follow:

Theorem 5.1 (successor state axiom for knowledge (SSAK))

$\models_{ES} \Box([a]\mathbf{K}(\varphi \leftrightarrow$
$SF(a) \wedge \mathbf{K}(Poss(a) \wedge SF(a) \rightarrow [a]\varphi \vee$
$\sim SF(a) \wedge \mathbf{K}(Poss(a) \wedge \sim SF(a) \rightarrow [a]\varphi).$

Proof For both directions of the equivalence we will only consider the case where $\sim SF(n)$ holds for an arbitrary action name n. The other case is completely analogous. To prove the only-if direction, let $e, w, z \models [a]\mathbf{K}(\varphi)$ for action a. Suppose $e, w, z \models \sim SF(a)$. It suffices to show that $e, w, z \models \mathbf{K}(Poss(a) \wedge \sim SF(a) \rightarrow [a]\varphi)$. So suppose $w' \simeq_z w, w' \in e, w'[Poss(a), z] = 1$, and $w'[SF(a), z] = 0$. Thus $w'[SF(a), z] = w[SF(a), z]$ and, hence, $w' \simeq_{z \cdot n} w$. Since $e, w, z \models [a]\mathbf{K}(\varphi')$ by assumption, $e, w', z, z \cdot a \models \varphi'$, from which $e, w', z \models [a]\varphi'$ follows.

Conversely, let $e, w, z \models \sim SF(a) \wedge \mathbf{K}(Poss(a) \wedge \sim SF(a) \rightarrow [a]\varphi'$. We need to show that $e, w, z \models [a]\mathbf{K}(\varphi')$, that is, $e, w, z \cdot a \models \mathbf{K}(\varphi')$. Let $w' \simeq_{z \cdot a} w$ and $w' \in e$. Then $w'[Poss(a), z] = 1$ and $w'[SF(a), z] = w[SF(a), z] = 0$ by assumption Hence e, $w', z \models Poss(a) \wedge \sim SF(a)$. Therefore, by assumption, $e, w', z \cdot a \models \varphi'$, from which $e, w, z \models [a]\mathbf{K}(\varphi')$ follows.

The Theorem 5.1 states that an agent already knows a conditional.

5.3 Rough Set and Decision Logic

Here, we briefly review rough set theory and decision logic of rough set. Decision logic is an application of rough sets to a deduction system and we describe how a deduction is represented with rough sets.

5.3.1 Rough Set

Rough set theory, proposed by Pawlak [24, 25], provides a theoretical basis of sets based on approximation concepts. A rough set can be seen as an approximation of a set. Therefore, rough sets are used for imprecise data handling. The semantic framework for general rough set theory is denoted by the notion of a knowledge base:

Definition 5.2 A knowledge base is a tuple $S = (U, \mathbf{R})$, where:

- U is a universe of objects;

- **R** is a set of equivalence relations on objects in U.

With each subset $X \subseteq U$ and an equivalence relation R, we associate two subsets:

Definition 5.3 Let $R \in \mathbf{R}$ be an equivalence relation of the knowledge base $S = (U, \mathbf{R})$, and X any subset of U. Then, the lower and upper approximations of X for R are defined as follows:

$$\underline{R}X =_{\text{def}} \bigcup\{Y \in U/R \mid Y \subseteq X\} = \{x \in U \mid [x]_{\mathrm{R}} \subseteq X\},$$

$$\overline{R}X =_{\text{def}} \bigcup\{Y \in U/R \mid Y \cap X \neq \emptyset\} = \{x \in U \mid [x]_{\mathrm{R}} \cap X \neq \emptyset\}.$$

Intuitively, $\underline{R}X$ is the set of all elements of U that can be certainly classified as elements of X in the knowledge R, and $\overline{R}X$ is the set of elements that can be possibly classified as elements of X in the knowledge R.

Then, we can define three types of sets:

Definition 5.4 If $S = (U, \mathbf{R})$ and $X \subseteq U$, then the *R-positive*, *R-negative*, and *R-boundary* regions of X with respect to R are defined as follows:

$$POS_R(X) = \underline{R}X$$

$$NEG_R(X) = U - \overline{R}X$$

$$BN_R(X) = \overline{R}X - \underline{R}X$$

If the positive and negative regions on a rough set are considered to correspond to the truth-value of a logical form, then the boundary region corresponds to ambiguity in deciding truth or falsity.

5.3.2 *Decision Logic*

In general, targets of rough set data analysis are described by table-style format called information tables. Formally, an information system is defined as a pair $S = (U, A)$, where U and A, are finite, nonempty sets called the universe, and the set of attributes, respectively.

With every attribute $a \in A$ we associate a set V_a, of its *values*, called the *domain* of a. Any subset B of A determines a binary relation $IND(B)$ on U, which is called an *indiscernibility relation*, and defined as follows:

$$IND(B) = \{(x, y) \in U^2 \mid \text{ for every } a \in B, a(x) = a(y)\},$$

where $a(x)$ denotes the value of attribute a for element x. $IND(B)$ is assumed as an equivalence relation.

The family of all equivalence classes of $IND(B)$ will be denoted by $U/IND(B)$, or simply by U/B where a partition is determined by B. A block of the partition U/B containing x will be denoted by $B(x)$.

If (x, y) belongs to $IND(B)$ we will say that x and y are *B-indiscernible* (indiscernible with respect to B). Equivalence classes of the relation $IND(B)$ are referred to as *B-elementary* sets.

If we distinguish in an information system two disjoint classes of attributes, called condition and decision attributes, respectively, then the system will be called a *decision table* and will be denoted by $S = (U, C, D)$, where C and D are disjoint sets of condition and decision attributes, respectively.

We review the foundations of rough set-based decision logic [24, 25]. Let $S = (U, A)$ be an information system. With every $B \subseteq A$ we associate a formal language as follows:

Definition 5.5 The set of formulas of the decision logic language $\mathscr{L}_{DL}$ is the smallest set satisfying the following conditions:

1. (a, v), or in short a_v, is an atomic formula, where the set of attribute constants is defined as $a \in A$ and the set of attribute value constants is $v \in V = \bigcup_{a \in A} V_a$,
2. If φ *and* ψ are formulas of the *DL*, then $\sim \varphi$, $\varphi \wedge \psi$, $\varphi \vee \psi$, $\varphi \to \psi$, and $\varphi \equiv \psi$ are formulas.

Formulas of $\mathscr{L}_{DL}$ are interpreted as subsets of objects consisting of a value v and an attribute a. An object $x \in U$ satisfies a formula φ in $S = (U, A)$, denoted $x \models_S \varphi$. The semantic relations of formulas are recursively defined as follows:

$$x \models_S (a, v) \text{ iff } a(x) = v,$$
$$x \models_S \sim \varphi \text{ iff } x \not\models_S \varphi,$$
$$x \models_S \varphi \vee \psi \text{ iff } x \models_S \varphi \text{ or } x \models_S \psi,$$
$$x \models_S \varphi \wedge \psi \text{ iff } x \models_S \varphi \text{ and } x \models_S \psi,$$
$$x \models_S \varphi \to \psi \text{ iff } x \models_S \sim \varphi \vee \psi,$$
$$x \models_S \varphi \equiv \psi \text{ iff } x \models_S \varphi \to \psi \text{ and } s \models_S \psi \to \varphi.$$

If $\varphi \in \mathscr{L}_{DL}$ is a formula then the set $|\varphi|_S$ defined as follows:

$$| \varphi |_S = \{x \in U \mid x \models_S \varphi\}$$

will be called the *meaning* of the formula φ in S.

Let φ be an atomic formula of $\mathscr{L}_{DL}$, $R \in C \cup D$ an equivalence relation, X any subset of U, and a valuation v of propositional variables.

Let L be a set of propositional constants of $\mathscr{L}_{DL}$ and $S : L \longrightarrow \{\mathbf{T}, \mathbf{F}\}$ be a valuation function. Let $\|\varphi\|_S$ be the interpretation of φ under S:

$$\|\varphi\|_S = \begin{cases} \mathbf{t} \text{ if } |\varphi|_S \subseteq POS_R(X) \\ \mathbf{f} \text{ if } |\varphi|_S \subseteq NEG_R(X) \end{cases},$$

where **t** represents the classical value true, and **f** represents the classical value false.

A decision table of Pawlak [24, 25] assumes consistency and excludes an inconsistent formula then it is denoted that the decision logic of Pawlak is based on classical bivalent logic.

The decision table and the decision logic can be utilized to construct a theory of ES by interpreting an attribute as a fluent. This application of the decision table and the decision logic is not described in this research and we describe in the future works.

5.3.3 Variable Precision Rough Set

The VPRS models proposed by Ziarko [28] are one extension of Pawlak's rough set theory, which provides a theoretical basis to treat probabilistic or inconsistent information in the framework of rough sets.

VPRS is based on the majority inclusion relation. Let $X, Y \subseteq U$ be any subsets of U. The majority inclusion relation is defined by the following measure $c(X, Y)$ of the relative degree of misclassification of X with respect to Y.

$$c(X, Y) =_{\text{def}} \begin{cases} 1 - \dfrac{|X \cap Y|}{|X|}, & \text{if } X \neq \emptyset, \\ 0, & \text{otherwise.} \end{cases}$$

where $|X|$ represents the cardinality of the set X. It is easy to confirm that $X \subseteq Y$ holds if and only if $cd(X, Y) = 0$.

Formally, the majority inclusion relation $\overset{\beta}{\subseteq}$ with a fixed precision $\beta \in [0, 0.5)$ is defined using the relative degree of misclassification as follows:

$$X \overset{\beta}{\subseteq} Y \text{ iff } c(X, Y) \leq \beta,$$

where the precision β provides the limit of permissible misclassification.

Let $X \subseteq U$ be any set of objects, R be an indiscernibility relation on U, and a degree $\beta \in [0, 0.5)$ be a precision. The β *– lower approximation* $\underline{R}_\beta(X)$ of X and the β *– upper approximation* $\overline{R}_\beta(X)$ of X by R are respectively defined as follows:

$$\underline{R}_\beta(X) =_{\text{def}} \left\{x \in U \mid c([x]_R, X) \leq \beta\right\},$$
$$\overline{R}_\beta(X) =_{\text{def}} \left\{x \in U \mid c([x]_R, X) < 1 - \beta\right\}.$$

As mentioned previously, the precision β represents the threshold degree of misclassification of elements in the equivalence class $[x]_R$ to the set X.

Thus, in VPRS, misclassification of elements is allowed if the ratio of misclassification is less than β. Note that the β − lower and β − upper approximations with $\beta = 0$ correspond to Pawlak's lower and upper approximations, respectively.

To follow the traditional notation of the theory of rough sets, the β-lower approximation will also be called the β-positive region of the set X and denoted alternatively as $POS_{R\beta}(X)$. The β-boundary region and the β-negative region of X are defined as follows:

$$BNR_{R\beta}(X) =_{\text{def}} \{x \in U \mid \beta < c([x]_R, X) < 1 - \beta\},$$
$$NEG_{R\beta}(X) =_{\text{def}} \{x \in U \mid c([x]_R, X) \geq 1 - \beta\}.$$

5.4 Zooming Reasoning

5.4.1 Kripke Model

Given a set of atomic sentences P, a language $\mathcal{L}_{ML}(\mathcal{P})$ for modal logic is formed from P using logical operators $\top$, $\bot$, $\neg$, $\wedge$, $\vee$, $\rightarrow$, $\leftrightarrow$, and two kinds of modal operators $\Box$ and $\Diamond$ as the least set of sentences generated by the following formation rules:

(1) $p \in \mathcal{P} \Rightarrow p \in \mathcal{L}_{ML}(\mathcal{P})$,
(2) $\top, \bot \in \mathcal{L}_{ML}(\mathcal{P})$,
(3) $p \in \mathcal{L}_{ML}(\mathcal{P}) \Rightarrow \neg p, \Box p, \Diamond p \in \mathcal{L}_{ML}(\mathcal{P})$,
(4) $p, q \in \mathcal{L}_{ML}(\mathcal{P}) \Rightarrow (p \wedge q), (p \vee q), (p \rightarrow q) \in \mathcal{L}_{ML}(\mathcal{P})$.

We will formulate our idea in the framework of possible world semantics, but, here, we do not use modal operators. Thus, following Chellas [7], we only assume the structure $\langle W, \ldots, v\rangle$, which we call a Kripke-style model in this paper, where W is a non-empty set of possible worlds, $v : \mathcal{P} \times W \rightarrow \{0, 1\}$, is a valuation, where 0 and 1 denote, respectively, false and true, and the ellipsis indicates the possibility of additional elements like a binary relation in the standard Kripke models.

Given a Kripke-style model $\mathcal{M} = \langle W, \ldots, v\rangle$, from a valuation v, a relationship among a model $\mathcal{M}$, a possible world w and an atomic sentence p, written $\mathcal{M}, w \models p$, is defined by

$$\mathcal{M}, w \models p \stackrel{\text{def}}{\Longleftrightarrow} v(p, v) = 1$$

and it is extended for every compound sentence in the usual inductive way. When we need to extend it to modal sentences, we must add some elements to the above ellipsis. Let

$$\|p\|^{\mathcal{M}} = \{w \in W \mid \mathcal{M}, w \models p\}$$

and thus

$$\mathcal{M}, w \models p \Leftrightarrow w \in \|p\|^{\mathcal{M}}.$$

5.4.2 Granularized Possible World and Zooming Reasoning

Let us consider a possible-worlds model $\mathcal{M} = \langle U, \ldots, v \rangle$, where U is a set of possible worlds and v is a binary valuation: $v : \mathcal{P} \times U \rightarrow \{0, 1\}$. When we need modal operators, we introduce either some accessibility relation in case of well-known Kripke models or by some neighborhood system in case of Scott-Montague models.

Let $\mathcal{P}$ be a set of atomic sentences and $\mathcal{L}_{\mathcal{P}}$ is the propositional language generated from $\mathcal{P}$ using a standard kind of set of connectives including modal operators in a usual way.

Also, let Γ be a subset of $\mathcal{L}_{\mathcal{P}}$ and let $\mathcal{P}_{\Gamma}$ be the set of atomic sentences which appears in each sentence in Γ is denoted as follows:

$$\mathcal{P}_{\Gamma} \equiv S_n(\Gamma) \cap \mathcal{P},$$

where $S_n(\Gamma) = \bigcup_{p \in \Gamma} S_n(p)$ is a subset of p [7].

Here, let R_{Γ} on U be an equivalence relation for any $p \in \mathcal{P}_{\Gamma}$, then we can define an equivalence relation R_{Γ} on U by

$$x \sim_{\Gamma} y \text{ iff } \forall p \in \mathcal{P}_{\Gamma}, \text{ where } V(p, x) = V(p, y).$$

which is called an *agreement relation* in [7].

Here, we regard the quotient set $U/\sim_{\Gamma}$ for Γ as a set of granularized possible worlds with respect to Γ, denoted $\tilde{U}_{\Gamma}$, as

$$\tilde{U}_{\Gamma} =_{\text{def}} U/R_{\Gamma} = \{[x]_{R_{\Gamma}} \mid x \in U\}.$$

This represents a set of elements of granularized possible worlds in Γ. The valuation of granularized possible worlds is

$$V_{\Gamma}(p, X) = 1 \text{ iff } p \in \cap X,$$

where $p \in \mathcal{P}_{\Gamma}$ and $X \in \tilde{U}_{\Gamma}$.

In addition, when an accessibility relation R on U and R' on $\tilde{U}_{\Gamma}$ meet the following conditions:

- if xRy, then $[x]_{\sim_{\Gamma}} R' [y]_{\sim_{\Gamma}}$,
- if $[x]_{\sim_{\Gamma}} R' [y]_{\sim_{\Gamma}}$, then for all $\Box p \in \Gamma$, $\mathcal{M}, x \models \Box p \Rightarrow \mathcal{M}, y \models p$,
- if $[x]_{\sim_{\Gamma}} R' [y]_{\sim_{\Gamma}}$, for all $\sim \Box \sim p \in \Gamma$, $\mathcal{M}, x \models \sim \Box \sim p \Rightarrow \mathcal{M}, y \models p$,

$\mathcal{M}_{\Gamma}^{R'} = \langle U_{\Gamma}, R', V_{\Gamma} \rangle$ is a *filtration* of a finite subset of $S_n(\Gamma)$.

Thus we have a granularization of a set of possible worlds. We also make granularization of a valuation as follows:

$$\tilde{v}_\Gamma : \mathscr{P} \times \tilde{U}_\Gamma \to 2^{\{0,1\}}.$$

Here, we obtain a definition of a valuation v as follows (cf. Murai et al. [18]:

$$\tilde{v}_\Gamma(p, \tilde{w}) = \begin{cases} \{1\}, \text{ if } v(p, w) = 1 \text{ and } v(p, w) \neq 0 \text{ for any } w \in \tilde{w}, \\ \{0\}, \text{ if } v(p, w) = 0 \text{ and } v(p, w) \neq 1 \text{ for any } w \in \tilde{w}, \\ \emptyset, \text{ if } v(p, w) \neq 1 \text{ and } v(p, w) \neq 0 \text{ for any } w \in \tilde{w}, \\ \{1, 0\}, \text{ if } v(p, w) = 1 \text{ and } v(p, w) = 0 \text{ for any } w \in \tilde{w}. \end{cases}$$

Originally the valuation is defined for three-valued interpretation in Murai [16]. In this chapter, we describe a granularization of a valuation based on four-valued semantics in the following section.

Now we have a granularized model for $\mathscr{M}$ with respect to Γ as $\tilde{\mathscr{M}}_\Gamma =_{def} \langle \tilde{U}_\Gamma, \ldots, \tilde{v}_\Gamma \rangle$. Based on this valuation, we can define the following partially defined relationship $\tilde{\mathscr{M}}, \tilde{w} \models p$.

For two finite subsets Γ', Γ such that $\Gamma' \subseteq \Gamma \subseteq \mathscr{L}_{\mathscr{P}}$, we have $R_\Gamma \subseteq R_{\Gamma'}$, then U_Γ is a refinement of $U_{\Gamma'}$. Now, we call a mapping $I_\Gamma^{\Gamma'} : \tilde{\mathscr{M}}_{\Gamma'} \to \tilde{\mathscr{M}}_\Gamma$ a *zooming in* from Γ' to Γ, and also call a mapping $O_{\Gamma'}^{\Gamma} : \tilde{\mathscr{M}}_\Gamma \to \tilde{\mathscr{M}}'_\Gamma$ a *zooming out* from Γ to Γ'.

For example, let $\{p\} = \Gamma' \subseteq \Gamma = \{p, q\}$, then we can make the following zooming in (Table 5.1) and zooming out (Table 5.2):

In monotonic reasoning case, the following relation is obtained.

$$\|p\|^{\mathscr{M}} \subseteq \underline{R}_{\{p\}}(\|q\|^{\mathscr{M}}) \subseteq \|q\|^{\mathscr{M}}.$$

Hence we have

Table 5.1 Truth table of Zooming In

$\tilde{U}_{\{p,q\}}$	p	q
$\|p\|^{\mathscr{M}} \cap \|q\|^{\mathscr{M}}$	1	1
$\|p\|^{\mathscr{M}} \cap (\|q\|^{\mathscr{M}})^C$	1	0
$(\|p\|^{\mathscr{M}})^C \cap \|q\|^{\mathscr{M}}$	0	1
$(\|p\|^{\mathscr{M}})^C \cap (\|q\|^{\mathscr{M}})^C$	0	0

Table 5.2 Truth table of Zooming Out

$\tilde{U}_{\{p\}}$	p	q
$\|p\|^{\mathscr{M}}$	1	$\{\emptyset, \{1, 0\}\}$
$(\|p\|^{\mathscr{M}})^C$	0	$\{\emptyset, \{1, 0\}\}$

$$\tilde{\mathscr{M}}, \|p\|^{\mathscr{M}} \models q \overset{\text{def}}{\Longleftrightarrow} \|p\|^{\mathscr{M}} \subseteq \underline{R}_{\{p\}}(\|q\|^{\mathscr{M}})$$

.

Secondly, the operation of zooming in from Γ' to Γ, where $\Gamma' \subseteq \Gamma$, increases the amount of information, and we can easily prove

$$\Gamma' \subseteq \Gamma \Rightarrow (\tilde{\mathscr{M}}_{\Gamma'} \models p \Rightarrow \tilde{\mathscr{M}}_{\Gamma} \models p),$$

which shows *monotonicity* of reasoning using the lower approximation.

5.5 Consequence Relation for Partial Semantics

5.5.1 Belnap's Four-Valued Logic

Belnap first claimed that an inference mechanism for a database should employ a certain four-valued logic in Belnap [5, 6]. The important point in Belnap's system is that we should properly deal with both incomplete and inconsistent information in databases.

To represent such information, we need a four-valued logic, since classical logic is not appropriate for the task. Belnap's four-valued semantics can in fact be viewed as an intuitive description of internal states of a computer.

In Belnap's four-valued logic **B4**, four kinds of truth-values are used from the set **4** = {**T**, **F**, **N**, **B**}.These truth-values can be interpreted in the context of a computer, namely **T** means just told True, **F** means just told False, **N** means told neither True nor False, and **B** means told both True and False. Intuitively, **N** can be interpreted as undefined and **B** as overdefined.

Belnap outlined a semantics for **B4** using the logical lattice **L4** (the order of $\leq_t$ in Fig. 5.1,[2] which has negation, conjunction and disjunction as logical connectives.

The ordering on **L4** we write as $a \leq b$; we write *meets* as $a \ \& \ b$, and *joins* as $a \vee b$. We can now use these logical operations on **L4** to induce a semantics for a language involving &, $\vee$, and $\sim$, in just the usual way.

Belnap's semantics uses a notion of *set-ups* mapping atomic formulas into **4** and a set-up can then be extended for any formula in **B4** in the following way:

$$s(A \ \& \ B) = s(A) \ \& \ s(B),$$
$$s(A \ \vee \ B) = s(A) \ \vee \ s(B),$$
$$s(\sim A) = \sim s(A).$$

[2] On **4** one can define two orderings: a truth ordering $\leq_t$ and an information ordering $\leq_i$ (often referred to as a knowledge ordering or approximation ordering, presented in the diagram below.

Fig. 5.1 Lattice **4**

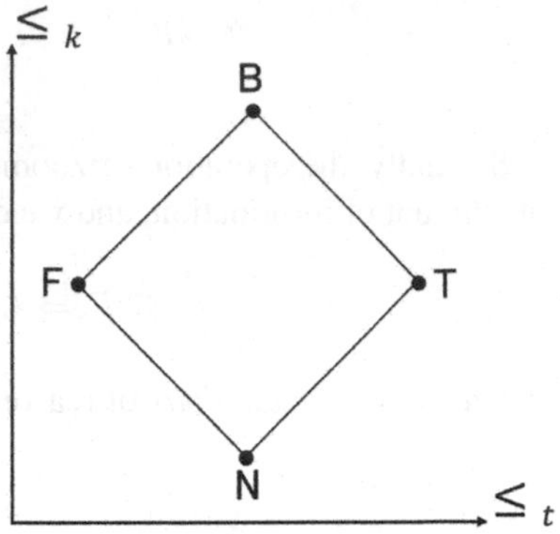

Belnap also defined a concept of entailments in **B4**. We say that A entails B just in case for each assignment of one of the four values to variables, the value of A does not exceed the value of B in **B4**, i.e., $s(A) \leq s(B)$ for each set-up s. Here, $\leq$ is transitive and defined as: $\mathbf{F} \leq \mathbf{B}, \mathbf{F} \leq \mathbf{N}, \mathbf{B} \leq \mathbf{T}, \mathbf{N} \leq \mathbf{T}$.

Belnap's four-valued logic in fact coincides with the system of *tautological entailments* due to Anderson and Belnap [3]. Belnap's logic **B4** is a paraconsistent logic capable of tolerating contradictions. Belnap also studied implications and quantifiers in **B4** in connection with question-answering systems.

5.5.2 Four-Valued Modal Logic

Here, we assume that an agreement relation $\simeq_{\langle\rangle}$ of ES as an equivalent relation R_Γ and define the agreement relation as a neighborhood relation of Scott-Montague model. In addition, we define a valuation of a formula and a granularized world as follows:

$$\tilde{v}_\Gamma : \mathscr{P} \times \tilde{U}_\Gamma \to 2^{\{0,1\}}.$$

The elements of the set **4** = {**T**,**F**,**N**,**B**} can be represented as subsets of the set {0, 1} of classical truth values:

$$\{\mathbf{T} = \{1\}, \mathbf{F} = \{0\}, \mathbf{N} = \emptyset, \mathbf{B} = \{0, 1\}.$$

Extended Belnap's four-valued logic **B4** is represented as the following tuple.

$$\mathbf{B4} := \langle\{\mathbf{4}, \wedge, \vee, \rightarrow, \bot, \sim, \{\mathbf{T}, \mathbf{B}\}\rangle.$$

We define an **N4**-model for Nelson's paraconsistent constructive logic **N4** as a triple $\mathscr{M} = \langle W, \leq, V\rangle$. Let $Prop$ be a set of propositions. Here, W is a non-empty set of possible worlds, $\leq$ is a preordering on W, and $V : Prop \times W \to \mathbf{4}$ is a valuation function satisfying the following monotonicity condition:

$$w \leq w' \Rightarrow V(p, w) \leq V(p, w').$$

Next, we represent the model and the interpretation for ES. For granularized possible worlds, we define the model of four-valued logic ES^4 (For simplicity we write ES) based on **B4**.

$$\tilde{\mathcal{M}} = \langle \tilde{W}, \simeq_{()}, \tilde{V} \rangle, \text{ where } \tilde{V} : Prop \times \tilde{W} \rightarrow \mathbf{4}.$$

We also define the model following two valuations of $\tilde{v}^+, \tilde{v}^-$ as follows:

$$\tilde{\mathcal{M}} = \langle \tilde{W}, \simeq_{()}, \tilde{v}^+, \tilde{v}^- \rangle$$

We define *verification* and *falsification* relations for ES [23],

$$\tilde{\mathcal{M}}, \tilde{w} \models^+ p \Leftrightarrow \tilde{w} \in \tilde{v}^+(p),$$
$$\tilde{\mathcal{M}}, \tilde{w} \models^- p \Leftrightarrow \tilde{w} \in \tilde{v}^-(p).$$

The valuation of a formula of ES is as follows:

$$1 \in \tilde{V}(p, \tilde{w}) \Leftrightarrow \tilde{w} \in \tilde{v}^+(p),$$
$$0 \in \tilde{V}(p, \tilde{w}) \Leftrightarrow \tilde{w} \in \tilde{v}^-(p).$$

The semantic relation of a formula is denoted as follows:

$$1 \in \tilde{V}(p, \tilde{w}) \Leftrightarrow \tilde{\mathcal{M}}, \tilde{w} \models^+ p,$$
$$0 \in \tilde{V}(p, \tilde{w}) \Leftrightarrow \tilde{\mathcal{M}}, \tilde{w} \models^- p.$$

Here, we define the valuation of formula in granularized worlds as follows:

Definition 5.6

$$\tilde{v}_\Gamma(p, \tilde{w}) = \begin{cases} \{1\}, \text{ if } \tilde{w} \in \tilde{v}^+(p), \\ \{0\}, \text{ if } \tilde{w} \in \tilde{v}^-(p), \\ \emptyset, \text{ if } \tilde{w} \notin \tilde{v}^+(p) \text{ and } \tilde{w} \notin \tilde{v}^-(p), \\ \{1, 0\}, \text{ if } \tilde{w} \in \tilde{v}^+(p) \text{ and } \tilde{w} \in \tilde{v}^-(p). \end{cases}.$$

5.5.3 Semantic Relation with Four-Valued Logic

Here, we define the following semantic relation, which is based on an agreement relation based on granular reasoning as partial semantics interpretation. To extend the bivalent semantics of ES to enable the partial interpretation, we assume a four-valued semantics for agreement relation for the epistemic interpretation of ES.

Let $\mathbf{4} = \{\mathbf{T}, \mathbf{F}, \mathbf{N}, \mathbf{B}\}$ be truth-value for the four-valued semantics of zooming reasoning system. To describe the property of partiality about zooming reasoning, we define the forcing relation based on four values as semantic relation.

Let the formula be p and define a valuation for the formula as follows:

$$V^4(p) = \begin{cases} \mathbf{T} \text{ iff } \tilde{\mathscr{M}}_\Gamma \models_v^+ p \text{ and } \tilde{\mathscr{M}}_\Gamma \not\models_v^- p, \\ \mathbf{F} \text{ iff } \tilde{\mathscr{M}}_\Gamma \not\models_v^+ p \text{ and } \tilde{\mathscr{M}}_\Gamma \models_v^- p, \\ \mathbf{N} \text{ iff } \tilde{\mathscr{M}}_\Gamma \not\models_v^+ p \text{ and } \tilde{\mathscr{M}}_\Gamma \not\models_v^- p, \\ \mathbf{B} \text{ iff } \tilde{\mathscr{M}}_\Gamma \models_v^+ p \text{ and } \tilde{\mathscr{M}}_\Gamma \models_v^- p. \end{cases}.$$

The definition of the valuation for the granularity in VPRS is also defined as the following valuation corresponding to the inclusion relation of each area of VPRS.

$$\tilde{v}_\Gamma(\mathrm{p}, \tilde{w}) = \begin{cases} \mathbf{T} \text{ iff } \|p\|^{\mathscr{M}} \subseteq POS_R(U/X), \\ \mathbf{F} \text{ iff } \|p\|^{\mathscr{M}} \subseteq NEG_R(U/X), \\ \mathbf{N} \text{ iff } \|p\|^{\mathscr{M}} \subseteq BNR_\beta(U/X), \\ \mathbf{B} \text{ iff } \|p\|^{\mathscr{M}} \subseteq POSR_\beta(U/X) \cap NEGR_\beta(U/X), \end{cases}.$$

where $\beta \in [0, 0.5)$. For **B**, it may also be interpreted as $BNR_\beta(U/X)$.

A semantic relation for model $\tilde{\mathscr{M}}_\Gamma$ is defined. The semantics such as true and false are denoted as $\models_v^+$ and $\models_v^-$ respectively.

Definition 5.7 The semantic relation for $\tilde{\mathscr{M}} \models_v^+ p$ and $\tilde{\mathscr{M}} \models_v^- p$ is defined as follows:

$\tilde{\mathscr{M}}_\Gamma, x \models_v^+ p$ iff $p \in \tilde{U}_\Gamma^+$,
$\tilde{\mathscr{M}}_\Gamma, x \models_v^- p$ iff $p \in \tilde{U}_\Gamma^-$,
$\tilde{\mathscr{M}}_\Gamma, x \models_v^+ \sim p$ iff $\mathscr{M}, x \models_v^- p$,
$\tilde{\mathscr{M}}_\Gamma, x \models_v^- \sim p$ iff $\mathscr{M}, x \models_v^+ p$,
$\tilde{\mathscr{M}}_\Gamma, x \models_v^+ p \vee q$ iff $\tilde{\mathscr{M}}_\Gamma, x \models_v^+ p$ or $\tilde{\mathscr{M}}_\Gamma \models_v^+ q$,
$\tilde{\mathscr{M}}_\Gamma, x \models_v^- p \vee q$ iff $\tilde{\mathscr{M}}_\Gamma, x \models_v^- p$ and $\tilde{\mathscr{M}}_\Gamma \models_v^- q$,
$\tilde{\mathscr{M}}_\Gamma, x \models_v^+ p \wedge q$ iff $\tilde{\mathscr{M}}_\Gamma, x \models_v^+ p$ and $\tilde{\mathscr{M}}_\Gamma \models_v^+ q$,
$\tilde{\mathscr{M}}_\Gamma, x \models_v^- p \wedge q$ iff $\tilde{\mathscr{M}}_\Gamma, x \models_v^- p$ or $\tilde{\mathscr{M}}_\Gamma \models_v^- q$,

A logical symbol $\sim$ means strong negation. Its interpretation is that the strong negation of a formula is true (false) iff the formula is false (true).

In possible world semantics using Kripke models, we use accessibility relations to interpret modal sentences. $\Box p$ is true at x if and only if p is true at every possible world y accessible from x, and $\Diamond p$ is true at x if and only if there is at least one possible world y accessible from x, and p is true at y. Formally, the interpretation of modal sentences is defined as follows:

$$\tilde{\mathscr{M}}_\Gamma, x \models_v^+ \Box p \text{ iff } \forall y \in U(xRy \Rightarrow \tilde{\mathscr{M}}, y \models_v^+ p),$$
$$\tilde{\mathscr{M}}_\Gamma, x \models_v^- \Box p \text{ iff } \exists y \in U(xRy \text{ and } \tilde{\mathscr{M}}, y \models_v^- p),$$
$$\tilde{\mathscr{M}}_\Gamma, x \models_v^+ \Diamond p \text{ iff } \exists y \in U(xRy \text{ and } \tilde{\mathscr{M}}, y \models_v^+ p),$$
$$\tilde{\mathscr{M}}_\Gamma, x \models_v^- \Diamond p \text{ iff } \forall y \in U(xRy \Rightarrow \tilde{\mathscr{M}}, y \models_v^- p).$$

For any sentence $p \in \mathscr{L}_{ML}(\mathscr{P})$, the truth set is the set of possible worlds in which p is true by the Kripke model $\mathscr{M}$, and the truth set is defined as follows:

$$\|p\|^{\mathscr{M}} \overset{\text{def}}{\Longleftrightarrow} \{x \in U | \mathscr{M}, x \models^+ p\}.$$
$$(\|p\|^{\mathscr{M}})^C \overset{\text{def}}{\Longleftrightarrow} \{x \in U | \mathscr{M}, x \models^- p\},$$
$$\tilde{\mathscr{M}}_\Gamma, x \models_v^+ \Box p \text{ iff } [x]_R \cap (\|p\|^{\tilde{\mathscr{M}}})^C = \emptyset,$$
$$\tilde{\mathscr{M}}_\Gamma, x \models_v^- \Box p \text{ iff } [x]_R \cap (\|p\|^{\tilde{\mathscr{M}}})^C \neq \emptyset,$$
$$\tilde{\mathscr{M}}_\Gamma, x \models_v^+ \Diamond p \text{ iff } [x]_R \cap \|p\|^{\tilde{\mathscr{M}}} \neq \emptyset,$$
$$\tilde{\mathscr{M}}_\Gamma, x \models_v^- \Diamond p \text{ iff } [x]_R \cap \|p\|^{\tilde{\mathscr{M}}} = \emptyset.$$

Therefore, when R is an equivalence relation, the following correspondence relationship holds between Pawlak's lower approximation and necessity, and that between Pawlak's upper approximation and possibility:

$$\underline{R}(\|p\|^{\tilde{\mathscr{M}}}) = \|\Box p\|^{\tilde{\mathscr{M}}},$$
$$\overline{R}(\|p\|^{\tilde{\mathscr{M}}}) = \|\Diamond p\|^{\tilde{\mathscr{M}}}.$$

A possible world model can be extended to a granularized possible model according to VPRS and measure-based model of Kudo et al. [12].

5.5.4 Consequence Relation and Sequent Calculus

Here we describe the consequences based on partial semantics. We define the consequence relation (C4) as in Akama and Nakayama [2] and Nakayama et al. [20–22] for the four-valued logic as follows;

$$\text{(C4) for all } V, \text{ if } V(Pre) \neq 0, \text{ then } (V(Cons) \neq 0.$$

(C4) can interpret undefined and inconsistent in addition to true and false and can be interpreted as a consequence relation for Belnap's four-valued logic **B4**. We also define the assignment function $V^{C4}(p)$ as follows:

As the semantics for C4, Belnap's **B4** is adopted here. We define the extension of the valuation function $V^{C4}(p)$ for an atomic formula p as follows:

$$\mathbf{T} =_{\text{def}} V^{C4}(p) = \{1\},$$
$$\mathbf{F} =_{\text{def}} V^{C4}(p) = \{0\},$$
$$\mathbf{N} =_{\text{def}} V^{C4}(p) = \{\},$$
$$\mathbf{B} =_{\text{def}} V^{C4}(p) = \{1, 0\}.$$

Next, we provide the sequent calculus GC4 as the consequence relation C4 to Belnap's **B4**. Let X and Y be set of formulas and let A and B be a formula.

Axiom

(ID)$X, A \vdash_{GC4} A, Y.$

Sequent Rules

(Weakening)$X \vdash_{GC4} Y \Rightarrow X, A \vdash_{GC4} A, Y.$

(Cut)$X, A \vdash_{GC4} Y$ and $X \vdash_{GC4} A, Y \Rightarrow X \vdash_{GC4} Y.$

$(\wedge R)X \vdash_{GC4} Y, A$ and $X \vdash_{GC4} Y, B \Rightarrow X \vdash_{GC4} Y, A \wedge B.$

$(\wedge L)X, A, B \vdash_{GC4} Y \Rightarrow X, A \wedge B \vdash_{GC4} Y.$

$(\vee R)X \vdash_{GC4} A, B, Y \Rightarrow X \vdash_{GC4} A \vee B, Y.$

$(\vee L)X, A \vdash_{GC4} Y$ and $X, B \vdash_{GC4} Y \Rightarrow X, A \vee B \vdash_{GC4} Y.$

$({\sim}{\sim} R)X \vdash_{GC4} A, Y \Rightarrow X \vdash_{GC4} {\sim}{\sim} A, Y.$

$({\sim}{\sim} L)X, A \vdash_{GC4} Y \Rightarrow X, {\sim}{\sim} A \vdash_{GC4} Y.$

$({\sim} \wedge R)X \vdash_{GC4} {\sim} A, {\sim} B, Y \Rightarrow X \vdash_{GC4} {\sim} (A \wedge B), Y.$

$({\sim} \wedge L)X, {\sim} A \vdash_{GC4} Y$ and $X, {\sim} B \vdash_{GC4} Y \Rightarrow X, {\sim} (A \wedge B) \vdash_{GC4} Y.$

$({\sim} \vee R)X \vdash_{GC4} {\sim} A, Y$ and $X \vdash_{GC4} {\sim} B, Y \Rightarrow X \vdash_{GC4} {\sim} (A \vee B), Y.$

$({\sim} \vee L)X, {\sim} A, {\sim} B \vdash_{GC4} Y \Rightarrow X, {\sim} (A \vee B) \vdash_{GC4} Y.$

Here, we add an axiom of *safety* in Dunn [11] to GC4.

$$(\text{Safety})X, A, {\sim} A \vdash_{GC4} B, {\sim} B, Y.$$

As a result, even if a contradiction is deduced, the system does not become *trivial*, and appropriate conclusions can be deduced.

We also introduce rules for *weak negation* and *weak implication* to guarantee the deduction theorem.

$$(\neg R) X, A \vdash_{GC4} Y \Rightarrow X \vdash_{GC4} \neg A, Y,$$
$$(\neg L) X \vdash_{GC4} A, Y \Rightarrow X, \neg A \vdash_{GC4} Y,$$
$$(\rightarrow_w R) X, A \vdash_{GC4} B, Y \Rightarrow X \vdash_{GC4} A \rightarrow_w B, Y,$$
$$(\rightarrow_w L) X, B \vdash_{GC4} Y \text{ and } X \vdash_{GC4} A, Y \ \Rightarrow\ X, A \rightarrow_w B \vdash_{GC4} Y.$$

The definition of a semantic relation of weak negation is provided below. Weak negation represents a lack of truth.

$$\|\neg p\|^{\mathcal{M}} = \begin{cases} \mathbf{T} \text{ if } \|\mathbf{p}\|^{\mathcal{M}} \neq \mathbf{T} \\ \mathbf{F} \text{ otherwise} \end{cases}$$

By the weak implication, modus ponens and deduction theorem hold for four-valued logic.

5.6 Zooming Reasoning as Action

5.6.1 Zooming Reasoning in ES

Here, we consider zooming reasoning as an action of ES and zooming out is regarded as an abstraction and zooming in is regarded as a refinement. Then these actions can be understood as an epistemic action. In the situation calculus, the value of the fluent is changed by executing the action.

For a propositional fluent, truth-value of fluent is updated by a result of the action. For sensing in the situation calculus, the agent recognizes the fluent value by sensing, sensing can be considered as a function for a valuation.

Since a granular reasoning copes with finite atomic sentences at each reasoning step, zooming reasoning is performed by granularizing of possible worlds according to a process of reasoning.

The important point of the zooming reasoning is the focus of reasoning steps. This is the contact point between the zooming reasoning and the epistemic situation calculus. From this point, we can overlay the concept of focus on the action logic in ES.

Now, we treat the zooming reasoning as an action of ES, called *zooming action*. In zooming action, the truth value of a sentence is determined with regard to the degree of worlds when the zooming reasoning is executed as an action and the granularity of the world is changed dynamically.

Zooming action is usual action and it takes propositional fluent as object and mapping which takes a set of worlds divided by partial atomic formulas. Therefore, zooming action regards the mapping of zooming in and zooming out as two actions.

In the context of epistemic situation calculus, zooming in and zooming out are captured in the following process. An agent X has an initial knowledge, and he recognizes the current situation.

When he causes an action for something, he recognizes a set of sentences denoted Γ, which is concerned with at a given time, and it is called *focus* at the time. When the agent X moves his viewpoint from the focus to another denoted Δ along time, he must reconstruct the set of granularized possible worlds. It is considered that Γ is the current focus, and Δ is the next or successor focus to which he will move.

More precisely, we describe the process of zooming reasoning according to the action theory of ES. First, an agent takes attention to the target situation that is a *scene* of scope denoted Ξ, which represents target fluent for action.

Second, the agent performs zooming action on the scope of the scene and takes focus Γ which is a set of fluents on the current reasoning process and successor focus at the next reasoning process.

We introduced two operations of zooming in and zooming out on sets of worlds; see Murai et al. [16]. For the purpose here, we need to extend the two operations of zooming in and out on models to describe propositional reasoning as a dynamic process.

A set Γ of formulas we are concerned with at a given time is called a *focus* and its elements *focal* ones at the time. When we move our viewpoint from one focus to another along time, we must reconstruct the set of granularized possible worlds. Let Γ be the current *focus* and let Δ be the next *focus* we will move to.

Definition 5.8 *(Zooming In and Zooming Out)*

1. When $\mathscr{P}_\Gamma \supseteq \mathscr{P}_\Delta$, we need granularization, which is a mapping $I^\Gamma_\Delta : U_\Gamma \to U_\Delta$, where, for any X in U_Γ, $\mathscr{I}^\Gamma_\Delta(X) =_{\text{def}} \{x \in U \mid x \cap \mathscr{P}_\Delta = (\cap X) \cap \mathscr{P}_\Delta$. We call this mapping a zooming out from Γ to Δ.
2. When $\mathscr{P}_\Gamma \subseteq \mathscr{P}_\Delta$, we need the inverse operation of granularization, which is a mapping $\mathscr{O}^\Gamma_\Delta : U^\Gamma \to 2^{U_\Delta}$, where $\mathscr{O}^\Gamma_\Delta(X) = \{Y \in U_\Delta \mid (\cap Y) \cap \mathscr{P}_\Gamma = \cap X\}$. We call this mapping zooming in from Γ to Δ.

5.6.2 Action Theory for Zooming Reasoning

In the context of situation calculus, we can interpret that zooming reasoning is corresponding to a basic action theory. Action precondition axioms (APA) are defined to describe the conditions for the zooming action that are going to be executed. An action of zooming reasoning requires two sets of formulas called focus, where one is the focus that is paid attention to in the first process of thinking.

We assume this first focus is included in the scene of the world, and also defined in APA as an initial knowledge. We assume that the scene is explicitly recognized by an agent. Then the focus transits from the first process of thinking to the successor process of thinking. We call the second focus on the *successor focus*.

In the context of situation calculus, the successor focus is defined in the successor state axioms (SSA). For focuses, the appropriate granularity is applied for the worlds of models according to the situation of reasoning processes.

A zooming action is the same as a normal action and it works for a propositional fluent, and like a zooming reasoning, it is a mapping that takes a set of worlds divided by atomic subexpressions as arguments.

Let Ξ be the *scene* for the world where an agent performs reasoning action. And then, let $\Gamma \subseteq \Xi \subseteq \mathscr{L}(\mathscr{P})$ be a finite set of sentences considered in the current reasoning step.

Here, we incorporate zooming operations in Definition 5.8 as zooming action to ES. As the first step of zooming reasoning, we define a *focus* as fluent literals at a given time. For each situation of the reasoning process, each *focus* is represented with both Γ for the current situation and Δ for the successor situation.

In addition, let $\Delta \subseteq \Xi \subseteq \mathscr{L}(\mathscr{P})$ be a finite set of sentences considered in the next reasoning step. The definition of the *scene* V_S relative to Ξ and the focus V_F relative to Γ (Δ) are as follows:

$$V_S(\Xi) =_{\text{def}} \mathscr{P} \cap sub(\Xi) = \mathscr{P}_\Xi,$$
$$V_F(\Gamma) =_{\text{def}} \mathscr{P} \cap sub(\Gamma) = \mathscr{P}_\Gamma.$$

where sub is the set of all subsentences of sentences defined as in Definition 5.9:

Definition 5.9 sub is defined as follows:

1. $sub(\mathbb{P}_n = \{\mathbb{P}_\ltimes\}$, for $n = 0, 1, 2, \ldots$
2. $sub(\top) = \{\top\}$.
3. $sub(\bot) = \{\bot\}$.
4. $sub(\neg A) = \{\neg A\} \cup sub(A)$.
5. $sub(A \wedge B) = \{A \wedge B\} \cup sub(A) \cup sub(B)$.
6. $sub(A \vee B) = \{A \vee B\} \cup sub(A) \cup sub(B)$.
7. $sub(A \rightarrow B) = \{A \rightarrow B\} \cup sub(A) \cup sub(B)$.
8. $sub(A \leftrightarrow B) = \{A \leftrightarrow B\} \cup sub(A) \cup sub(B)$.

Now, we provide the basic action theory for zooming action. The action preconditions axioms and the successor state axioms for *zooming in* and *zooming out* are defined as follows:

Basic Action Theory:

Action Precondition Axiom (APA):

$$\Sigma_{pre} = \{Poss(a) \Leftrightarrow \Gamma \cup V_F(\Gamma)\}, \text{ where } a = [\mathscr{O}^\Gamma_\Delta] \text{ or } [\mathscr{I}^\Gamma_\Delta].$$

Successor State Axiom (SSA):

$$\Sigma_{post} = \{[a]\Delta \Leftrightarrow a = b \wedge (\Delta \cup V_F(\Delta))\}, \text{ where } b = [\mathscr{O}^\Gamma_\Delta] \text{ or } [\mathscr{I}^\Gamma_\Delta].$$

Sensed Fluent:

$\Sigma_{sense} = \{SF(a) \Leftrightarrow a = b \Rightarrow \varphi\}$, where $b = [\mathscr{O}_\Delta^\Gamma]$ or $[\mathscr{I}_\Delta^\Gamma]$.

For APA and SSA, the *zooming out* action is required to meet the following condition:

$\Gamma \supseteq \Delta$ and $\Gamma \cap \Delta \neq \emptyset$,

and the *zooming in* action is required to meet the following condition:

$\Gamma \subseteq \Delta$ and $\Gamma \cap \Delta \neq \emptyset$.

In the situation calculus, SSA is defined for each fluent with positive effect and negative effect.

5.6.3 Semantics for Zooming Reasoning in ES

Let Γ be the current focus and let Δ be the next or successor focus we will move to. For the zooming out, when $V_F(\Gamma) \supseteq V_F(\Delta)$, a granularization is required for mapping $\mathscr{O}_\Delta^\Gamma : U_\Gamma \rightarrow U_\Delta$, where, for any X in U_Γ,

$$\mathscr{O}_\Delta^\Gamma(X) =_{\text{def}} \{x \in U | x \cap V_F(\Delta) = (\cap X) \cap V_F(\Delta)\}.$$

For the zooming in, when $V_F(\Gamma) \subseteq V_F(\Delta)$, an inverse operation of granularization is required for mapping $\mathscr{I}_\Delta^\Gamma : U_\Gamma \rightarrow 2^{U_\Delta}$, where, for any X in U_Γ,

$$\mathscr{I}_\Delta^\Gamma(X) =_{\text{def}} \{Y \in U_\Delta | (\cap Y) \cap V_F(\Gamma) = (\cap X)\}.$$

For the zooming reasoning, an atomic sentence p in Ξ and an elementary world at APA meets the following condition:

$$\Sigma \models p \Leftrightarrow V_\Xi(p, x) = 1 \text{ and } p \in x.$$

When a binary relation R is given on U_Ξ, we have Kripke model: $\mathscr{M}_\Xi = \langle U_\Xi, R, V_\Xi \rangle$ for the zooming reasoning.

$V_F(\Gamma)$ (or $V_F(\Delta)$) is the set of atomic formulas of fluent in Γ and $p \in \Gamma$ and $\Gamma \models p$ means that, for any possible world $w \in W$, if $\mathscr{M}, w \models q$ for any sentence $q \in \Gamma$ then $\mathscr{M}, w \models p$.

Here, we assume a neighbourhood model for an agreement relation. Using truth valuation function v, we construct agreement relation $\sim_{\mathscr{P}_\Gamma}$ ($\sim_{\mathscr{P}_\Delta}$) based on focus $\mathscr{P}_\Gamma = V_F(\Gamma)$ as follows:

$$x \sim_{\mathscr{P}_\Gamma} y \Leftrightarrow v(p, x) = v(p, y), \text{ where } \forall p \in V_F(\Gamma).$$

Agreement relation $\sim_{\mathscr{P}_\Gamma}$ therefore induces quotient set $\tilde{W} =_{\text{def}} W / \sim_{\mathscr{P}_\Gamma}$ of W, i.e., the set of all equivalence classes of possible worlds in W. We also construct truth valuation $\tilde{V}_{\mathscr{P}_\Gamma}$ for equivalence classes of possible worlds $\tilde{x} =_{\text{def}} [x]_{\mathscr{P}_\Gamma} \in \tilde{W}$. An agreement relation for granularized world is represented as follows:

$$\tilde{W} : (\mathscr{P} \cup Poss(a) \cup SF(a)) \times A^*) \rightarrow 2^{\{0,1\}}, \text{ where } a \in A.$$

Then, we also get the following relation for a granularized world.

1. $\tilde{w}' \simeq_{()} \tilde{w}$, for every $\tilde{w}'$ and $\tilde{w}$
2. $\tilde{w}' \simeq_{z,a} \tilde{w}$ iff $\tilde{w}' \simeq_z \tilde{w}$ and $\tilde{w}'([z]SF(a)) = \tilde{w}([z]SF(a))$, where $a = \mathscr{O}_{\Gamma}^{\Delta}$.

The definition of a valuation $V_{\mathscr{P}_\Gamma}$ relative to a focus Γ is as follows:

$$\tilde{V}_{\mathscr{P}_\Gamma} : V_F(\Gamma) \times \tilde{W} \rightarrow 2^{\{0,1\}}.$$

The valuation function for focus on granular worlds for Γ is defined as follows:

Definition 5.10

$$\tilde{V}_{\mathscr{P}_\Gamma}(p, \tilde{w}) = \begin{cases} \{1\} \text{ if } p \in \mathscr{P}_\Sigma \text{ and } \tilde{w} \in \tilde{W}, \\ \{0\} \text{ if } p \in \mathscr{P}_\Sigma \text{ and } \tilde{w} \notin \tilde{W}, \\ \{1, 0\} \text{ if } p \in \mathscr{P}_\Sigma \text{ and } (\tilde{w} \in \tilde{W} \text{ and } \tilde{w} \notin \tilde{W}), \\ \emptyset \text{ if } p \notin \mathscr{P}_\Sigma. \end{cases}$$

It is also clear the following semantic relation:

If for all $p \in \mathscr{P}_\Gamma$ then $V_{\mathscr{P}_\Gamma}(p, \tilde{w}) = \{1\}$,
If for all $p \notin \mathscr{P}_\Gamma$ then $V_{\mathscr{P}_\Gamma}(p, \tilde{w}) = \{0\}$.

In zooming action, the valuation is determined on the granularized world. We denote the semantic relation with *verification* and *falsification* as the valuation based on the four-valued extended ES.

$$\langle e, w, z\rangle \models^+ [a]p \text{ iff } \langle e, w, z \cdot a\rangle \models^+ p,$$
$$\langle e, w, z\rangle \models^- [a]p \text{ iff } \langle e, w, z \cdot a\rangle \models^- p.$$

The valuation of formulas is as follows:

$$1 \in \tilde{V}(p, \tilde{w}) \Leftrightarrow \langle e, w, z \cdot a\rangle \models^+ p,$$
$$0 \in \tilde{V}(p, \tilde{w}) \Leftrightarrow \langle e, w, z \cdot a\rangle \models^- p.$$

In zooming action, which is as same as usual action, the following relation holds.

$$\langle e, w, z\rangle \models_{ES} [\mathscr{I}_{\Delta}^{\Gamma}]p \Leftrightarrow \langle e, w, z \cdot \mathscr{I}_{\Delta}^{\Gamma}\rangle \models_{ES} p,$$
$$\langle e, w, z\rangle \models_{ES} [\mathscr{O}_{\Delta}^{\Gamma}]p \Leftrightarrow \langle e, w, z \cdot \mathscr{O}_{\Delta}^{\Gamma}\rangle \models_{ES} p.$$

The interpretation of zooming action in granularized epistemic world is represented as follows:

$$\langle e, w, z\rangle \models_{ES} [\mathscr{I}_{\Delta}^{\Gamma}]p \Leftrightarrow \|p\|^{\tilde{\mathscr{M}}} \subseteq \underline{R}_{\{\Gamma\}}(\|p\|^{\mathscr{M}}),$$
$$\langle e, w, z\rangle \models_{ES} [\mathscr{O}_{\Delta}^{\Gamma}]p \Leftrightarrow \|p\|^{\tilde{\mathscr{M}}} \subseteq \overline{R}_{\{\Delta\}}(\|p\|^{\mathscr{M}}).$$

Zooming out:
An agent performs the reasoning at the abstraction level with an action of zooming out. Zooming out action is an abstraction in the epistemic world, and performs the zooming out reasoning at a larger granularized world which arrow vagueness.

If $\{p\} = e_2 \subseteq e_1 = \{p, q\}$, where $E = e_1 \cup e_2$ then it is shown that $[\mathscr{O}_{e_2}^{e_1}](p \rightarrow q)$ holds as follows:

$$\langle e_1, e_2, w, z\rangle \models_{ES} [\mathcal{O}^{e_1}_{e_2}](p \rightarrow q).$$

This can be read as "generally" or "typically" $p \rightarrow q$, and also represented semantically as follows:

$$\not\models^{-}_{ES} [\mathcal{O}^{e_1}_{e_2}]p \text{ or } \models^{+}_{ES} [\mathcal{O}^{e_1}_{e_2}]q.$$

As the truth value of q is interpreted as $\tilde{v}_{\Gamma}(q, \tilde{w}) = \{0, 1\}$, above reasoning holds. In addition, this interpretation is deduced from Table 5.1.

Furthermore, let p be antecedent, and q consequent. Since *designated value*, i.e., $\{\mathbf{T}, \mathbf{B}\}$ holds at the valuation of e_1 and e_2 in zooming out, this reasoning is valid on the four-valued logic.

Zooming in:
An agent performs the reasoning at the refinement level with an action of zooming in. Zooming in action is refinement in the epistemic world, and performs the reasoning at a relative subdivided granularized world which enables concrete reasoning. If $\{p\} = e_2 \subseteq e_1 = \{p, q\}$, where $E = e_1 \cup e_2$, then it is shown that $[\mathcal{O}^{e_1}_{e_2}](p \rightarrow q)$ holds:

Therefore, it is shown that $[\mathcal{I}^{e_2}_{e_1}](p \rightarrow q)$ holds as follows:

$$\langle e_2, e_1, w, z\rangle \models_{ES} [\mathcal{I}^{e_2}_{e_1}](p \rightarrow q).$$

This can be read as "concretely" or "specifically" $p \rightarrow q$, and also represented semantically as follows:

$$\models^{+}_{ES} [\mathcal{I}^{e_2}_{e_1}]p \text{ and } \models^{+}_{ES} [\mathcal{I}^{e_2}_{e_1}]q,$$

and

$$\models^{-}_{ES} [\mathcal{I}^{e_2}_{e_1}]q \text{ and } \models^{-}_{ES} [\mathcal{I}^{e_2}_{e_1}]p.$$

Therefore this meets the validity of reasoning in four-valued logic (cf. Dunn [11]).

Figure 5.2 represents the relationship of epistemic worlds for abstraction and refinement. We can capture this relation as visibility relation in the epistemic world of an agent.

Example 5.1 We assume the following reasoning.

Tweety is a bird; (Most) birds fly.

(Typically) Tweety flies.

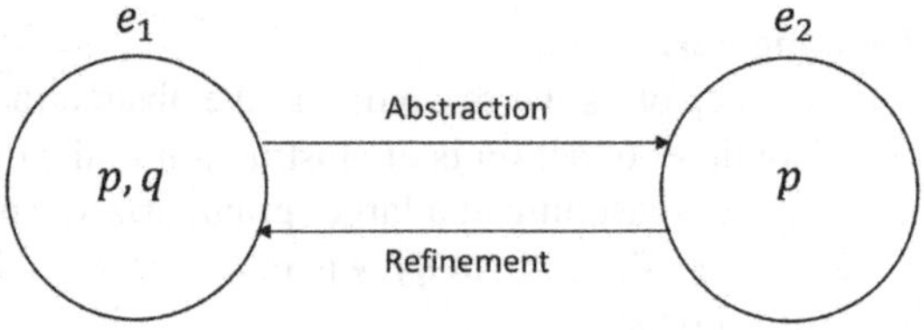

Fig. 5.2 Zooming action in epistemic world

while,

Tweety is a penguin; Penguin does not fly.
―――――――――――――――――――――
Tweety does not fly.

Each variable is defined as follows:

p: bird, q: fly, r: penguin, s: tweety

Zooming out reasoning:

An agent performs zooming out reasoning for epistemic world from $e_1 = \{p, q\}$ to $e_2 = \{p\}$. A granularized world is constructed by granularization with $\{p\}$. As the initial knowledge, it is assumed that $s \in \|r\|^{\mathcal{M}}$, $\|r\|^{\mathcal{M}} \subseteq \|p\|^{\mathcal{M}}$

By the zooming out from e_1 to e_2, reasoning for (typically) bird → fly is represented as follows:

$$\langle e_1, e_2, w, z \rangle \models^{+}_{ES} [\mathcal{O}^{e_1}_{e_2}](p \to q).$$

As a semantic relation, it is represented as follows:

$$\langle e_1, e_2, w, z \rangle \not\models^{-}_{ES} [\mathcal{O}^{e_1}_{e_2}]p \text{ or } \langle e_1, e_2, w, z \rangle \models^{+}_{ES} [\mathcal{O}^{e_1}_{e_2}]q.$$

Then, p and q are represented using Ziarko's inclusion relation [28] as follows:

$$\|p\|^{\mathcal{M}} \cap \underline{R}^{\beta}_{\{bird\}}(\|q\|^{\mathcal{M}}) \neq \emptyset.$$

The truth value of q is evaluated as $\tilde{v}_\Gamma(p, \tilde{w}) = \{0, 1\}$, then the validity is held in the deduction of four-valued logic.

Zooming in reasoning:

Suppose a proposition q is added into epistemic world $e_1 ande_2$, and $\mathcal{M}, r \models \sim q$ is added into the initial knowledge. Then the following interpretation is obtained. $\|r\|^{\mathcal{M}} \subseteq (\|q\|^{\mathcal{M}})^C$.

Zooming in reasoning is executed for the epistemic world from $e_2 = \{p, r\}$ to $e_1 = \{p, q, r\}$ as follows: At first, the following possible world is obtained: $\tilde{U}_\Gamma = \{\|r\|^{\mathcal{M}}, \|p\|^{\mathcal{M}} \cap (\|r\|^{\mathcal{M}})^C, (\|p\|^{\mathcal{M}})^C \cap (\|r\|^{\mathcal{M}})^C\}$.

Here, the zooming in from e_2 to e_1, we lead to the following:

$$\langle e_1, e_2, w, z \rangle \models_{ES} [\mathcal{I}^{e_2}_{e_1}](s \to \sim q).$$

This can be evaluated as follows: $\tilde{v}_\Gamma(p, \tilde{w}) = \{1\}$, $\tilde{v}_\Gamma(q, \tilde{w}) = \{0\}$, $\tilde{v}_\Gamma(r, \tilde{w}) = \{1\}$, and represented as follows:

$$\langle e_2, e_1, w, z \rangle \models^{+}_{ES} [\mathcal{I}^{e_2}_{e_1}]s \text{ and}$$
$$\langle e_2, e_1, w, z \rangle \models^{-}_{ES} [\mathcal{I}^{e_2}_{e_1}]q.$$

This means that the antecedent is true and the consequent false, then it leads that the reasoning cannot be held in the concretized world by adding an information r.

The situation of the following example is from Dennett [9].

Example 5.2 There was a robot, named RE by its creators. Its designers arranged for it to learn that its spare battery was put in a room with a time bomb. RE formulated a plan to rescue its battery. There was a wagon in the room, and the battery was on the wagon, and RE plans to pull out the battery being removed from the room.

First, the robot performs zooming out to grasp what to do to rescue its battery and understood to pull out all things on their wagon. Actually, this causes the bomb went off.

Next, he performs zooming in to realize what not to do to rescue its battery by refinement of its situation. Then he realizes the bomb next to the battery on the same wagon is explosive.
Abbreviation: $batt$: $battery$, $ptbl$: $portable$, rsc : $rescue$, exp : $explosion$.

The initial database Σ_0, first focus Γ and second focus Δ have the following relation: $\Sigma_0 \supseteq \Gamma \supseteq \Delta$
Zooming Out:

$\Sigma_0 = \{batt, bomb, on\_wagon,\ batt \wedge on\_wagon \rightarrow ptbl,\ bomb \wedge on\_wagon \rightarrow ptbl,\ batt \wedge safe,\ bomb \wedge \neg safe,\ \neg safe \rightarrow \neg ptbl\}$.

$\Sigma_{pre} = \{\Box Poss(\mathcal{O}^{\Gamma}_{\Delta}) \equiv \Sigma_0 \cup \Gamma\}$, where $\Gamma = \{batt, bomb, on\_wagon\}$.

$\Sigma_{post} = \{\Box[a]on\_wagon \equiv a = \mathcal{O}^{\Gamma}_{\Delta} \wedge on\_wagon\}$, where $\Delta = \{on\_wagon\}$.

$\Sigma_{sense} = \{\Box SF(a) \equiv a = \mathcal{O}^{\Gamma}_{\Delta} \wedge (batt \wedge wagon) \rightarrow ptbl\ \wedge\ (bomb \wedge wagon) \rightarrow ptbl.$

In zooming out, the battery and bomb are on the same wagon. The agent does not mind whether the bomb beside the battery in the mind of the focus to the battery. The sensing gets the agent to decide to move out the wagon with the battery and the bomb at the same time.
Zooming In:

$\Sigma_0 = \{batt, bomb, on\_wagon,\ batt \wedge on\_wagon \rightarrow ptbl,\ bomb \wedge on\_wagon \rightarrow ptbl,\ batt \wedge safe, bomb \wedge \neg safe,\ \neg safe \rightarrow \neg ptbl\}$.

$\Sigma_{pre} = \{\Box Poss(\mathcal{I}^{\Gamma}_{\Delta}) \equiv \Sigma_0 \cup \Gamma\}$, where $\Gamma = \{on\_wagon\}$.

$\Sigma_{post} = \{[a]exp \equiv a = \mathcal{I}^{\Gamma}_{\Delta} \wedge ptbl \wedge bomb \vee [a]rsq \equiv a = \mathcal{I}^{\Gamma}_{\Delta} \wedge ptbl \wedge batt\}$, where $\Delta = \{on\_wagon, batt, bomb, exp\}$.

$\Sigma_{sense} = \{\Box SF(a) \equiv a = \mathcal{I}^{\Gamma}_{\Delta} \wedge (batt \wedge wagon) \rightarrow ptbl\ \wedge\ (bomb \wedge wagon) \rightarrow \neg ptbl\}$.

In zooming in, the battery and bomb are distinguished on the wagon. The agent recognizes the bomb the wagon, and also, the bomb is not safe to bring out with the battery. The sensing gets the agent to deduce not to move out the wagon with the battery and the bomb at the same time.

5.7 Conclusion

In this chapter, we studied a basic framework in which we incorporate the zooming reasoning based on granular computing into the epistemic situation calculus. In ES, for the action can be regarded as the modal operator, then we treat zooming reasoning as a modal operator, and this enables us to provide the interpretation for the dynamic change of granularity for a world.

In addition, we provided the interpretation for the non-monotonic reasoning using zooming reasoning by the semantic relation of four-valued logic. We can construct the deduction system to deduce the relevant result without corruption as a logical system when the inconsistent result is deduced.

References

1. Akama, S., Murai, T., Kudo, Y.: Reasoning with Rough Sets. Springer, Heidelberg (2018)
2. Akama, S., Nakayama, Y.: Consequence relations in DRT. In: Proc. of the 15th International Conference on Computational Linguistics (COLING 1994) 2, pp. 1114–1117 (1994)
3. Anderson, A., Belnap, N.: Entailment: The Logic of Relevance and Necessity I. Princeton University Press, Princeton (1976)
4. Banihashemi, B., Giacomo, G., Lesperance, Y.: Abstraction in situation calculus action theories. In: Proc. of the 31th AAAI Conference on Artificial Intelligence, pp. 1048–1055 (2017)
5. Belnap, N.D.: A useful four-valued logic. In: Dunn, J.M., Epstein, G. (eds.) Modern Uses of Multi-Valued Logic, pp. 8–37. Reidel, Dordrecht (1977)
6. Belnap, N.D.: How a computer should think. In: Ryle, G. (ed.) Contemporary Aspects of Philosophy, pp. 30–55. Oriel Press (1977)
7. Chellas, B.: Modal Logic: An Introduction. Cambridge University Press, Cambridge (1981)
8. Demolombe, R.: Belief change: From situation calculus to modal logic. J. Appl. Non-Classical Logics **13**, 187–198 (2003)
9. Dennett, D.: Cognitive wheels: The frame problem of AI. In: Hookway, C. (ed.) Minds, Machines and Evolution, pp. 129–151. Cambridge University Press, Cambridge (1984)
10. Ditmarsch, H., Herzig, A., Lima, T.: From situation calculus to dynamic epistemic logic. J. Logic Comput. **21**, 179–204 (2011)
11. Dunn, J.: Partiality and its dual. Studia Logica **66**, 5–40 (2000)
12. Kudo, Y., Murai, T., Akama, S.: A granularity-based framework of deduction, induction, and abduction. Int. J. Approximate Reasoning **50**, 1215–1226 (2009)
13. Lakemeyer, G., Levesque, H.: A semantic characterization of a useful fragment of the situation calculus with knowledge. Artif. Intell. **175**, 142–164 (2011)
14. Levesque, H., Lakemeyer, G.: The Logic of Knowledge Bases. MIT Press, Cambridge, Mass (2001)
15. McCarthy, J., Hayes, P.: Some philosophical problems from the standpoint of Artificial Intelligence. In: Meltzer, B., Michie, D. (eds.) Machine Intelligence 4, pp. 463–502. Edinburgh University Press, Edinburgh (1969)
16. Murai, T., Sato, Y., Resconib, G., Nakata, M.: Granular reasoning using zooming in & out Part 1. Propositional reasoning (Extended Abstract). In: Proc. of the International Workshop on Rough Sets, Fuzzy Sets, Data Mining, and Granular-Soft Computing RSFDGrC 2003, pp. 421–424 (2003)
17. Murai, T., Sato, Y., Resconib, G., Nakata, M.: Granular reasoning using zooming in & out Part 2. Aristotle's categorical syllogism. Electron. Notes Theor. Comput. Sci. **82**, 186–197 (2003)
18. Murai, T., Sanada, M., Kudo, Y.M., Kudo, M.: A Note on Ziarko's variable precision rough Set model and nonmonotonic Reasoning. In: Rough Sets and Current Trends in Computing RSCTC 2004. Lecture Notes in Computer Science, vol. 3066, pp. 103–108 (2004)
19. Murai, T., Kudo, Y., Akama, S.: Paraconsistency, Chellas's conditional logics, and association rules. In: Akama, S. (ed.) Towards Paraconsistent Engineering, pp. 179–196. Springer, Heidelberg (2016)

20. Nakayama, Y., Akama, S., Murai, T.: Four-valued semantics for granular reasoning towards frame problem. In: Proc. of SCIS & ISIS, pp. 37–42 (2018)
21. Nakayama, Y., Akama, S., Murai, T.: Four-valued tableau calculi for decision logic of rough set. In: Knowledge-Based and Intelligent Information & Engineering Systems: Proc. of the 22nd International Conference, KES-2018, pp. 383–392 (2018)
22. Nakayama, Y., Akama, S., Murai, T.: Deduction system for decision logic based on many-valued logics. Int. J. Adv. Intell. Syst. **11**, 115–126 (2018)
23. Odintsov, S., Wansing, H.: Modal logics with Belnapian truth values. J. Appl. Non-Classical Logics **20**, 279–304 (2010)
24. Pawlak, Z.: Rough sets. Int. J. Comput. Inf. Sci. **11**, 341–356 (1982)
25. Pawlak, Z.: Rough Sets: Theoretical Aspects of Reasoning about Data. Kluwer, Dordrecht (1991)
26. Reiter, R.: Knowledge in Action: Logical Foundations for Specifying and Implementing Dynamical Systems. MIT Press. Mass, Cambridge (2001)
27. Schwering, C., Lakemeyer, G.: Projection in the epistemic situation calculus with belief conditionals. In: AAAI'15 Proc. of the 29th AAAI Conference on Artificial Intelligence, pp. 1583–1589 (2015)
28. Ziarko, W.: Variable precision rough set model. J. Comput. Syst. Sci. **46**, 39–59 (1993)

Chapter 6
Discussion

Abstract In Chap. 6, we discuss some aspects of the Frame Problem in the context of granular reasoning. We describe the essential concern behind the Frame Problem and discuss our point of view for the Frame Problem.

6.1 The Frame Problem

The Frame Problem is one of the most important problems in Artificial Intelligence (AI), raised by McCarthy and Hayes [39]. Since it is closely related to common-sense reasoning, we need to give some words for the problem.

6.1.1 What is the Frame Problem?

Although there may be many possible definitions of the Frame Problem, it is essentially the problem of describing in a computationally reasonable manner what properties persist and what properties change as actions are performed. More simply, the problem consists in specifying what does not change when some events occur.

The importance of the Frame Problem is that we cannot list for every possible-action and for every possible state of the world how that action changes the truth of individual facts. Thus, the Frame Problem is closely related to the so-called *non-monotonic reasoning*.

The Frame Problem was a hot topic in AI in the 1980's. In 1987, the workshop on The Frame Problem in Artificial Intelligence was held in Lawrence, Kansas, USA; see Brown [6].

According to Brown, who is the organizer of the workshop, there are the following approaches to the Frame Problem.

S. Akama et al., *Epistemic Situation Calculus Based on Granular Computing*,
Intelligent Systems Reference Library 239,
https://doi.org/10.1007/978-3-031-28551-6_6

- The circumscription-based approaches
- The modal logic-based approaches
- The syntactic consistency-based approaches
- The pragmatic a4pproaches

The circumscription-based approaches use the so-called *circumscription* proposed by McCarthy [37], employing the idea of minimizing the undetermined instances of a predicate by adding an axiom scheme.

This version of circumscription is a *predicate circumscription*. Later, McCarthy generalized predicate circumscription to *formula circumscription* for dealing with a formula; see McCarthy [38].

The modal logic approaches are based on *modal logic*, interpreting that propositions that hold in some intermediate future state obtained by an action to the current state K are those entailed by P where P holds in K and is logically possible with the physical laws and the obvious results of the action.

Schwind [52] developed a modal logic ZK for the representation of actions and resolved the Frame Problem. ZK can be characterized semantically by a Kripke structure with two state transition relations.

Brown [7] also developed a modal logic Z for the representation of knowledge. His logic is stronger than modal logic S5 and can serve as a correct theory of non-monotonic reasoning.

Akama [1] proposed the solution to the Frame problem unifying Veltmen's *data logic* (cf. Veltman [58]) and a theory of presuppositions. Akama's theory can logically cope with non-monotonicity in common-sense reasoning in the framework of a version of (partial) modal logic.

The syntactic consistency approaches, which are similar to the modal logic approaches, employ the proof-theoretic concept of consistency. Goebel and Goodwin [21] presented an approach to the planning problem by an approximate reasoning system based on *theory formation*.

Ginsberg and Smith [22, 23] proposed the solution to the Frame Problem in two papers. The former outlined a possible worlds approach to the Frame Problem which involves a single model of the world that is updated when actions are performed.

The latter dealt with the qualification problem by avoiding difficulties with each action we associate a set of domain constraints potentially blocking the action. Their approach has some similarities with modal logic-based approaches.

The pragmatic approaches suggest more pragmatic solution to the Frame Problem. For instance, Champeaux [9] proposed a fragment of intensional logic to solve the Frame Problem.

Rolston [46] described a tense-logic based approach to the Frame Problem by using the concept of applying *tense logic* to the problem. Rolston also presented several examples in the tense-logic programming system called *Chronolog*.

Haas [25] discussed *domain-specific frame axioms*. He argued that the universal frame axiom cannot work in a domain allowing incomplete descriptions of situations and pointed out the need of the domain-specific frame axioms.

Elgot-Drapkin et al. [13] proposed a framework of *real-time reasoning* applied to default reasoning. They discussed applications of their real-time reasoning to the Frame problem.

6.1.2 The Frame Problem in Logic

McCarthy and Hayes used mathematical logic to formalizing common-sense reasoning. They formalized situation calculus in the framework of first-order logic and they faced the Frame Problem. The difficulty lies in the inability of expressing non-monotonicity in common-sense reasoning.

Because the outline of situation calculus has been already given in Chap. 2, we here give a general discussion on situation calculus. McCarthy and Hayes developed situation calculus to solve the Frame Problem. It is based on the assumption about the complete state of the universe at an instant of time, where the laws of motion determine, given a situation, all future situations.

It is therefore not surprising to use (many-sorted) first-order logic as the basis for situation calculus. They distinguish global and local situations, where the former concerns the complete state of the universe at a time and the latter only specific parts thereof. This means that local situations may be tractable even if global situations should be intractable.

McCarthy and Hayes also thought *laws of nature* and *causal kinds* are related to common-sense reasoning. For instance, complete states of the universe determine subsequent complete states according to laws of motion which belong to laws of nature. Causal kinds suggest that complete state of the universe determine subsequent complete state of the universe by some *causality*,

However, after an action, some properties persist but other properties may change. And we need to logically specify this feature, namely this is to solve the Fame Problem. From a different perspective, it is the same as the representation of non-monotonicity.

Although many people attempted to provide the solution to the Frame Problem based on situation calculus mentioned above, The interest of research has moved to formalizing non-monotonicity in a logical setting.

In classical logic, the set of conclusions that can be drawn from a set of formulae always increases with the addition of new formulae. The fact led McCarthy and others to non-monotonic logics. Similar criticism of classical logic as a basis for common-sense reasoning due to its lack of non-monotonicity may be found in Minsky [43].

Non-monotonic logics could be regarded to be promising for the Frame Problem, but they could not solve another problem, i.e. the *Yale shooting Problem* of Hanks and McDermott [26].

The Yale Shooting Problem challenges the previous approaches to the Frame Problem. There is a sequence of situations. In the initial situation a gun is not loaded and the target is alive. In the next situation the gun is loaded. Eventually, a shot is fired, perhaps with fatal consequences.

In this scenario there are two "fluents", alive and loaded, and two actions, load and shoot. Being loaded and being alive are inert propositions. However, if they are true at a given moment, they will be true at the next moment unless some action such as shoot has taken place.

Later, the Yale Shooting Problem was solved by several different fashions. Many earlier non-monotonic formalisms were revised by accommodating to the Yale Shooting Problem. Note that most papers in the 1987 Workshop was devoted to the solution to it.

Many people claimed that logic-based approaches to common-sense reasoning are not suitable, but this may not be conclusive, In fact, some non-monotonic theories solve the Yale shooting problem. And other approaches mentioned above can resolve the Frame Problem.

We know that since the early 1980's a number of non-monotonic logics have been proposed in the literature. They include: McDermott and Doyle's *non-monotonic logic* (McDermott and Doyle [40, 41]), Reiter's *default logic* (Reiter [49]) and Moore's *autoepistemic logic* (Moore [44, 45]). We do not give a systemic survey of non-monotonic logics here. For the interested reader, see Lukasiewicz [34].

Some workers explored a general theory for non-monotonic logics using formal devices of mathematical logic. One of the ideas is to investigate non-monotonic consequence relations proof-theoretically as well as model-theoretically.

Gabbay [17] first tried such an attempt. Makinson [35] generalized Gabbay's formalism with semantic observations. Shoham [56] proposed to analyze non-monotonic logics in the so-called *preferential logic* based on a model-theoretic fashion.

We can find significant developments in the work on non-monotonic consequence relations and related topics. For instance, Kraus, Lehmann and Magidor [31] and Lehmann and Magidor [33] obtained various formal results. For a general survey, see Makinson [36].

There are in fact many other approaches to non-monotonic reasonings. For example, Veltman [59] proposed *update semantics* for default reasoning. Boutilier [3, 4] developed conditional logics for non-monotonic reasoning.

It is also noticed that we can find the development of *logic programming*, originally due to Kowalski [27, 28], which plays a crucial role for automating non-monotonic reasoning.

It is famous that negation implemented in logic programming languages like *Prolog*, relies on *negation as failure* (NAF) (cf. Clark [8]) and that NAF is a non-monotonic rule. Clark gave an operational semantics for NAF and interpreted NAF in classical logic using the concept of *completion*.

The semantics for logic programming has been extensively studied in connection with non-monotonic logics and *logic database* (deductive database). Van Emden and Kowalski [57] showed the semantics for Horn clause logic programs.

They developed three types of semantics, i.e. model-theoretic semantics, fixpoint semantics and operational semantics and proved their equivalences. Later, Apt and van Emden [2] extended the semantics for logic programs with NAF.

Logic databases formalize databases based on mathematical logic. It is thus possible to describe logic databases using logic programming. Gallaire and Minker [18] contains several approaches to logic databases. Papers in Minker [42] explore logic databases in the context of logic programming.

Many semantics for logic programs with NAF emerged including *three-valued semantics* (cf. Fitting [15], Kunen [32]) and *stable model semantics* (cf. Gelfond and Lifschitz [19]). For an overview on negation in logic programming, consult Shepherdson [55].

In the areas of logic programming, we can see two important extensions for non-monotonic reasoning with new kind of negation in addition to NAF to express explicit negative information. Gelfond and Lifschitz [20] proposed *logic programs with classical negation*. Kowalski and Sadri [29] developed a similar system called *logic programs with exceptions*.

These two semantics are based on the so-called the *answer set semantics*, generalizing the stable model semantics. These extensions obviously enhance the expressive power of logic programming for handling common-sense reasoning.

We here return to the Frame Problem. Kowalski and Sergot [30] proposed a logic-based formalism called *event calculus* for representing and reasoning about events. Event calculus can also solve the Frame Problem similar to situation calculus. In addition, as event calculus is based on Horn logic, we can automate reasoning about actions in logic programming languages.

By the 2000's, the Frame Problem did not become serious even in logic-based approaches. In other words, it becomes an important test for AI. This reveals that logic-based approaches to AI are still alive. In relation to the Frame Problem, Reiter [48] completely discussed reasoning about actions with logical theory and its implementation.

Shanahan [53] also concerns the history of research on the Frame Problem in the period. For a general survey on the Frame Problem; see Shanahan [54]. It can solve the Frame Problem similar to situation calculus. Since event calculus is formalized by Horn logic, it can be implemented by logic programming languages.

6.1.3 The Epistemological Frame Problem

It is well known that the Frame Problem was also discussed by some philosophers. Such an attempt is called the *epistemological Frame Problem*, and is of special importance to AI.

Notably, Dennett [10] seriously took up the Frame Problem from a philosophical viewpoint. Dennett [11] also further the argument in a robot scenario which will be discussed at length in the next section.

Dennett's understanding of the Frame Problem is not logical but epistemological. Thus, we should regard the Frame Problem as an epistemological problem, in that a cognitive creature can update his beliefs when it performs an action so that they remain roughly faithful to the world. Fodor [16] interpreted the Frame Problem similarly.

Although the Frame Problem has been adequately resolved by logic-based AI approaches, the epistemological Frame Problem survives for idealized rationality. Since the work of Fodor, modularity has been an explicit theme of cognitive research.

In fact, Fodor's claim suggests that the human cognitive system possesses a number of important subsystems that are modular. If the classical computational approach to the Frame Problem neglects modularity, it may be necessary to understand such modular subsystems.

If we understand the Frame Problem in the robot world, the importance of the epistemological Frame Problem would be clearer. As said before, one of the roots of the Frame Problem in logic-based approaches lies in computational limitations in a computer.

In the epistemological Frame Problem discussed by philosophers, the aspect roughly corresponds to the issue of the computational theory of mind. Although AI researchers could of course neglect the epistemological Frame Problem for their purposes, it raises other features of the Frame Problem in AI capable of constructing more ideal AI systems. In this sense, it is useful to discuss the Frame Problem by philosophical as well as AI viewpoints.

6.2 Re-Examination of the Frame Problem in the Context of Granular Reasoning

Axiomatizing Actions in the Situation Calculus

We can observe actions in that they have preconditions: requirements that must be satisfied whenever they can be executed in the current situation. A predicate symbol $Poss$ is introduced; $Poss(a, s)$ means that it is possible to perform the action a in that state of the world resulting from performing the sequence of actions s.
Here are some examples:

- If it is possible for a robot r to pick up an object x in situation s, then the robot is not holding any object, it is next to x, and x is not heavy:
  $Poss(pickup(r, x), s) \to [(\forall z)\neg holding(r, z, s)] \wedge heavy(x) \wedge nextTo(r, x, s)$.
- Whenever it is possible for a robot to repair an object, then the object must be broken, and there must be glue available:
  $Poss(repair(r, x), s) \to hasGlue(r, s) \wedge broken(x, s)$.

The next feature of dynamic worlds that must be described are the causal laws–how actions affect the values of fluents. These are specified by so-called *effect axioms*. The following are some examples:

- The effect on the relational fluent broken of a robot dropping a fragile object:
  $fragile(x, s) \to broken(x, do(drop(r, x), s))$.

This is the situation calculus way of saying that dropping a fragile object causes it to become broken; in the current situation s, if x is fragile, then in that successor situation $do(drop(r, x), s)$ resulting from performing the action $drop(r, x)$ in s, x will be broken.

- A robot repairing an object causes it not to be broken: $\neg broken(x, do(repair(r, x), s))$.
- Painting an object with colour c: $colour(x, do(paint(x, c), s)) = c$.

The Qualification Problem for Actions

With only the above axioms, nothing interesting can be proved about when an action is possible. For example, here are some preconditions for the action pickup:

$$Poss(pickup(r, x), s) \rightarrow [(\forall z)\neg holding(r, z, s)] \wedge \neg(x) \wedge nextTo(r, x, s).$$

The reason nothing interesting follows from this is clear; we can never infer when a pickup is possible. We can try reversing the implication:

$$[(\forall z)\neg holding(r, z, s)] \wedge \neg heavy(x) \wedge nextTo(r, x, s) \rightarrow Poss(pickup(r, x), s).$$

Now we can indeed infer when a pickup is possible, but unfortunately, this sentence is false. We also need, in the antecedent of the implication:

$$\neg gluedToFloor(x, s) \wedge \neg armsTied(r, s) \wedge \neg hitByTenTonTruck(r, s) \wedge \cdots$$

i.e., we need to specify all the qualifications that must be true in order for a pickup to be possible! For the sake of argument, imagine succeeding in enumerating all the qualifications for pickup.

Suppose the only facts known to us about a particular robot R, object A, and situation S are:

$$[(\forall z)\neg holding(R, z, S)] \wedge \neg heavy(a) \wedge nextTo(R, A, S).$$

We still cannot infer $Poss(pickup(R, A), S)$ because we are not given that the above qualifications are true! Intuitively, here is what we want: When given only that the "important" qualifications are true:

$$[(\forall z)\neg holding(R, z, S)] \wedge \neg heavy(a) \wedge nextTo(R, A, S).$$

and if we do not know that any of the "minor" qualifications–$\neg gluedToFloor(A, S)$, $\neg hitBvTenTonTruck(R, S)$–are true, infer $Poss(pickup(R, A), S)$.

But if we happen pen to know that anyone of the minor qualifications is false, this will block the inference of $Poss(pickup(R, A), S)$. Historically, this has been seen to be a problem peculiar to reasoning about actions, but this is not really the case.

Consider the following fact about *birds*, which has nothing to do with reasoning about actions:

$$bird(x) \land \neg penguin(x) \land \neg ostrich(x) \land \neg pekingDuck(x) \land \cdots \rightarrow flies(x).$$

But given only the fact $bird(Tweety)$, we want intuitively to infer $flies(Tweety)$. Formally, this is the same problem as action qualifications:

- The important qualification is $bird(x)$.
- The minor qualifications are: $\neg penguin(x), \neg ostrich(x), \cdots$

This is the classical example of the need for *non-monotonic reasoning* in Artificial Intelligence (AI). For the moment, it is sufficient to recognize that the qualification problem for actions is an instance of a much more general problem, and that there is no obvious way to address it.

We shall adopt the following (admittedly idealized) approach: Assume that for each action $A(x)$, there is an axiom of the form

$$Poss(A((x), s) \equiv \Pi_A(\mathbf{x}, s),$$

where $Poss(A((x), s)$ is a first-order formula with free variables x, s that does not mention the function symbol do. We shall call these action precondition axioms. For example:

$$Poss(pickup(r, x), s) \equiv [(\forall z)\neg holding(r, z, s)] \land \neg heavy(x) \land nextTo(r, x, s).$$

In other words, we choose to ignore all the "minor" qualifications, in favour of necessary and sufficient conditions defining when an action can be performed.

6.2.1 Solution to the Frame Problem with the Situation Calculus

There is another well known problem associated with axiomatizing dynamic worlds; axioms other than effect axioms are required. These are called *frame axioms*, and they specify the action invariants of the domain, i.e., those fluents unaffected by the performance of an action.

For example, the following is a positive frame axiom, declaring that the action of robot r' painting object x' with colour c has no effect on robot r holding object x:

$$holding(r, x, s) \rightarrow holding(r, x, do(paint(r', x'), s)).$$

Here is a negative frame axiom for not breaking things:

$$\neg broken(x, s) \wedge [x \neq y \vee \neg fragile(x, s)] \rightarrow \neg broken(x, do(drop(r, y), s)).$$

Notice that these frame axioms are truths about the world, and therefore must be included in any formal description of the dynamics of the world.

The problem is that there will be a vast number of such axioms because only relatively few actions will affect the value of a given fluent. All other actions leave the fluent invariant.

For example, an object's colour remains unchanged after picking something up, opening a door, turning on a light, electing a new prime minister of Canada, etc.

Since, empirically in the real world, most actions have no effect on a given fluent, we can expect of the order of $2 \times \mathscr{A} \times \mathscr{F}$ frame axioms, where $\mathscr{A}$ is the number of actions, and $\mathscr{F}$ the number of fluents.

These observations lead to what is called the Frame Problem:

1. The axiomatizer must think of, and write down, all these quadratically many frame axioms. In a setting with 100 actions and 100 fluents, this involves roughly 20, 000 frame axioms.
2. The implementation must somehow reason efficiently in the presence of so many axioms.

Suppose the person responsible for axiomatizing an application domain has specified all the causal laws for that domain. More precisely, she has succeeded in writing down all the effect axioms, i.e. for each relational fluent $\mathscr{F}$ and each action $\mathscr{A}$ that causes F's truth value to change, axioms of the form:

$$R(\mathbf{x}, s) \rightarrow (\neg)F(\mathbf{x}, do(A, s)),$$

and for each functional fluent f and each action A that can cause f's value to change, axioms of the form:

$$R(\mathbf{x}, y, s) \rightarrow f(\mathbf{x}, do(A, s)) = y.$$

Here, R is a first-order formula specifying the contextual conditions under which the action A will have its specified effect on F and f. There are no restrictions on R, except that it must refer only to the current situation s. Later, we shall be more precise about the syntactic form of these effect axioms.

A solution to the Frame Problem is a systematic procedure for generating, from these effect axioms, all the frame axioms. If possible, we also want a parsimonious representation for these frame axioms (because in their simplest form, there are too many of them).

Reiter [49] described the reason to use frame axioms in the solution to the Frame Problem as follows:

- *Modularity*. As new actions and/or fluents are added to the application domain, the axiomatizer need only add new effect axioms for these. The frame axioms will be automatically compiled from these (and the old frame axioms suitably modified to reflect these new effect axioms).

- *Accuracy*. There can be no accidental omission of frame axioms.

Frame Axioms: Pednault's Proposal

Pednault [47] proposed ADL (Action Description Language) for planning, with a conditional operator. Therefore, ADL is useful to describe frame axioms. The example illustrates a general pattern. Assume given a set of positive and negative effect axioms (one for each action $A(\mathbf{y})$ and fluent $F(\mathbf{x}, s)$:

$$\varepsilon_F^+(\mathbf{x}, \mathbf{y}, s) \rightarrow F(\mathbf{x}, do(A(\mathbf{y}, s)), \tag{6.1a}$$

$$\varepsilon_F^-(\mathbf{x}, \mathbf{y}, s) \rightarrow \neg F(\mathbf{x}, do(A(\mathbf{y}, s)), \tag{6.1b}$$

Axioms 6.1a and 6.1b specify all the causal laws relating the action A and the fluent F. With this completeness assumption, we can reason as follows: Suppose that both $F(x, s)$ and $\neg F(\mathbf{x}, do(A(\mathbf{y}), s))$ hold. Then because F was true in situation s, action A must have caused it to become false.

By the completeness assumption, the only way A could cause F to become false is if $\varepsilon_F^-(\mathbf{x}, \mathbf{y}, s)$ were true. This can be expressed axiomatically by:

$$F(\mathbf{x}, s) \wedge \neg F(\mathbf{x}, do(A(\mathbf{y}), s)) \rightarrow \varepsilon_F^-(\mathbf{x}, \mathbf{y}, s).$$

Summary of Pednault's proposal:

- For deterministic actions, it provides a systematic (and easily and efficiently implementable) mechanism.
- generating frame axioms from effect axioms. But it does not provide a parsimonious representation of the frame axioms.

Frame Axioms: The Haas/Schubert Proposal

Schubert, elaborating on a proposal of Haas, argues in favor of what he calls *explanation closure axioms* for representing the usual frame axioms ([25, 50]).

Example 6.1 Consider a robot r that is holding an object x in situation s, but is not holding it in the next situation: Both $holding(r, x, s)$ and $\neg holding(r, x, do(a, s))$ are true. How can we explain the fact that holding ceases to be true?

If we assume that the only way this can happen is if the robot r put down or dropped x, we can express this with the explanation closure axiom:

$$holding(r, x, s) \wedge \neg holding(r, x, do(a, s)) \rightarrow a = putDown(r, s) \vee a = drop(r, s).$$

This says that all actions other than putDown(r, x) and drop(r, x) leave holding invariant, which is the standard form of a frame axiom (actually, a set of frame axioms, one for each action distinct from put Down and drop).

In general, an explanation closure axiom has one of the two forms:

$$F(\mathbf{x}, s) \wedge \neg F(\mathbf{x}, do(a, s)) \rightarrow \alpha_F(\mathbf{x}, a, s),$$
$$\neg F(\mathbf{x}, s) \wedge \neg F(\mathbf{x}, do(a, s)) \rightarrow \beta_F(\mathbf{x}, a, s),$$

In these, the action variable a is universally quantified. These say that if ever the fluent F changes truth value, then α_F or β_F provides an exhaustive explanation for that change.

As before, to see how explanation closure axioms function like frame axioms, rewrite them in the logically equivalent form:

$$F(\mathbf{x}, s) \wedge \neg\alpha_F(\mathbf{x}, a, s) \rightarrow F(\mathbf{x}, do(a, s)),$$

and

$$\neg F(\mathbf{x}, s) \wedge \neg\alpha_F(\mathbf{x}, a, s) \rightarrow \neg F(\mathbf{x}, do(a, s)).$$

Schubert argues that explanation closure axioms are independent of the effect axioms, and it is the axiomatizer's responsibility to provide them. Like the effect axioms, these are domain-dependent.

In particular, Schubert claimes that they cannot be obtained from the effect axioms by any kind of systematic transformation. Thus, Schubert and Pednault entertain conflicting intuitions about the origins of frame axioms.

6.2.2 Dennett's Robot Scenario

The earliest reference to the Frame Problem from a philosophical point of view was Dennett's "Cognitive Wheels: The Frame Problem of AI" published in 1987; see Dennett [11].

Dennett [11] appropriates the name "frame problem" to cover more than just the narrow technical problem defined by McCarthy and Hayes [39]. According to Dennett, the problem arises from our widely held assumptions about the nature of intelligence and it is "a new, deep epistemological problem brought by the artificial intelligence, and still being an open problem".

Here is a robot scenario from Dennett [11]:

> Once upon a time there was a robot, named R1 (...). One day its designers arranged for it to learn that its spare battery (...) was locked in a room with a time bomb set to go off soon. R1 located the room (...) and formulated the plan to rescue the battery.
>
> There was a wagon in the room, and the battery was on the wagon, and R1 hypothesized that a certain action (...) would result in the battery being removed from the room. Straightway it acted, and did succeed in getting the battery out of the room before the bomb went off.

> Unfortunately, however, the bomb was also on the wagon. R1 knew that the bomb was on the wagon, but did not realize that pulling the wagon would bring the bomb out along with the battery.

R1 did not pay attention to the ramifications of his actions. Thus, the designers decided that R1 must be redesigned to be able to recognize all the implications of his actions.

> Our next robot must be made to recognize not just the intended implications of its acts, but also the implications about their *side-effects*, by deducing these implications from the descriptions it uses in formulating its plans.'
>
> They called their next model, the robot-deducer, R1D1. They placed R1D1 in much the same predicament that R1 succumbed to, and (...) it began, as designed, to consider the implications of such a course of action. It had just finished deducing that pulling the wagon out of the room would not change the color of the room's walls, and was embarking on a proof of the further implication that pulling the wagon would cause its wheels to turn more revolutions that there were wheels in the wagon–when the bomb exploded.

This time, R1D1 was equipped with all necessary knowledge to perform the action correctly, yet it failed to reach the proper conclusion in reasonable amount of time. This, one might say, is the computational aspect of the frame problem.

The obvious way to avoid it is to appeal to the notion of relevance. Only certain properties are relevant in the context of any given action and we can confine the deduction only to those.

> Back to the drawing board. 'We must teach it the difference between relevant implications and irrelevant implications,' said the designers, 'and teach it to ignore the irrelevant ones.' So they developed a method of tagging implications as either relevant or irrelevant to the project at hand, and installed the method in their next model, the robot-relevant-deducer, or R2D1 for short.

For R2D1 case, the definition of relevancy is needed as a strict discussion. So we cannot recognize and determine what is relevant in this situation. It seems to be impossible to see that specifying what propositions are relevant to what context.

The following is pointed out in Gryz [24]. Contexts are not independent of each another, one needs to appeal to a larger context to determine the significance of elements in a narrower context (for example, to recognize two dots as eyes in a picture, one must have already recognized the context as a face).

But then, as Dreyfus described: "if each context can be recognized only in terms of features selected as relevant and interpreted in a broader context, the AI worker is faced with a regress of contexts (Dreyfus [12]).

Dreyfus [14] claims that this "extreme version of the frame problem" is no less a consequence of the Cartesian assumptions of classical AI and cognitive science than its less demanding relatives [14].

He advances the view that a suitably Heideggerian account of mind is the basis for dissolving the frame problem here too, and that our "background familiarity with how things in the world behave" is sufficient, in such cases, to allow us to "step back and figure out what is relevant and how". Dreyfus does not explain how, given the holistic, open-ended, context-sensitive character of relevance, this figuring-out is achieved.

From the situation calculus point of view, The assumption of the initial knowledge of the robot is as follows:

- The battery is needed to be rescued.
- The bomb is going to explode and it is dangerous.

We also need to assume initial knowledge like that the number of a wagon is one and the battery and the bomb are on the wagon, and he knows various behaviors for manipulation.

By this limited situation, we tentatively proposed reasoning about the safety with respect to the bomb using zooming reasoning.

6.3 Conclusion

The solution from the situation calculus is limited to the way to describe precondition axiom and effect axiom for action and fluent and does not cover how to treat the world around an agent.

For example, Reiter's approach for the solution to the Frame Problem is to create a mechanism which deduces the frame axiom automatically. Such an approach to the Frame Problem is based on classical logic in the situation calculus and it cannot resolve the complete description of the frame axiom for the world. It is impossible to describe all states of the world in limited physical time for an agent.

As the narrow approach, it is said that technical problem is largely solved, and recent discussion has tended to focus less on matters of interpretation and more on the implications of the wider science [5].

Gryz [24] surveyed the Frame Problem from the original approach of McCarthy and Hayes [39] to the typical philosophical point of view, e.g. Dennett [11]. Gryz pointed out that the Frame Problem is not solved with a logic approach. He described that a more likely explanation is that after so many failed attempts researchers lost faith that the problem can be solved by logic.

He also describes Brooks' argument. Brooks took to heart Dreyfus's arguments and attempted to build robots following a different methodological paradigm (Brooks [5]) that avoids running into the Frame Problem. The success of this project has been rather limited, but perhaps the only way to overcome the frame problem is to avoid it rather than solving it.

Ever since it was first pointed out by McCarthy and Hayes [39], the Frame Problem has remained a large concern in the research field of artificial intelligence and philosophy.

As many philosophers agree, the Frame Problem is concerned with how an agent may efficiently filter out irrelevant information in the process of problem-solving. Despite many attempts, no completely satisfactory solution has been obtained.

Zooming Reasoning for the Frame Problem:

Our point of view of the Frame Problem is based on the partiality of knowledge of an agent. The approach to Frame Problem from the partiality of epistemic state, this dismissed the concern of the Frame Problem.

Zooming reasoning can perform the reasoning as presuppose of limited knowledge, and can make the granularized rough world that holds for a general interpretation with abstraction.

This leads to the release of an agent to consider complete information about the actual world. Instead, an agent does not know all the information about the world but the partial information.

This epistemic perspective enables make of some kind of unconscious worlds instead of frame axioms. Consider again the following default reasoning example: *birds*, which has nothing to do with reasoning about actions:

$$bird(x) \land \neg penguin(x) \land \neg ostrich(x) \land \neg pekingDuck(x) \land \cdots \rightarrow flies(x).$$

We do not need to consider these exceptions to describe the *general bird*. We can interpret *most bird* using the action of zooming out.

As for the essential and philosophical frame problem raised by Dennett [11], this remains still an open problem in AI. Our approach of zooming action as an epistemic action is one of the challenges and this action can be controlled by an agent intentionally. Therefore, an agent chooses and focuses the target intentionally as its goal to the plan which an agent to perform.

This does not elucidate the problem raised in Dennett's robot R2D1 to distinct relevant implications and irrelevant implications. This problem will still be left to our concerns about how to describe the relevancy of ontology.

References

1. Akama, S.: Presupposition and frame problem in knowledge bases. In: Brown, F.M. (ed.) The Frame Problem in Artificial Intelligence, pp. 193–202. Morgan Kaufmann, Los Altos (1987)
2. Apt, K., van Emden, M.: Contributions to the theory of logic programming. J. ACM **29**, 841–863 (1982)
3. Boutilier, C.: Conditional logics of normality as modal systems. In: Proc. of AAAI'90, pp. 1134–1139 (1990)
4. Boutilier, C.: Conditional logics for default reasoning and belief revision. Ph.D. Thesis, University of Toronto (1992)
5. Brooks, R.: Intelligence without representation. Artif. Intell. **47**, 139–159 (1991)
6. Brown, F.M. (ed.): The Frame Problem in Artificial Intelligence. Morgan Kaufmann, Los Altos (1987)
7. Brown, F.M.: A modal logic for the representation of knowledge. In: Brown, F.M. (ed.) The Frame Problem in Artificial Intelligence, pp. 135–157. Morgan Kaufmann, Los Altos (1987)
8. Clark, K.: Negation as failure. In: Gallaire, H., Minker, J. (eds.) Logic and Data Bases, pp. 293–322. Plenum Press, New York (1978)
9. de Champeaux, D.: Unframing the frame problem. In: The Frame Problem in Artificial Intelligence, pp. 311–318. Morgan Kaufmann, Los Altos (1987)

10. Dennett, D.: Brainstorms. MIT Press, Cambridge, Mass (1978)
11. Dennett, D.: Cognitive wheels: The frame problem of AI. In: Hookway, C. (ed.) Minds, Machines and Evolution, pp. 129–151. Cambridge University Press, Cambridge (1984)
12. Dreyfus, H.: What Computers Still Can't Do. MIT Press, Cambridge, Mass (1992)
13. Elgot-Drapkin, J., Miller, M., Perlis, D.: Life on a desert island; Ongoing work on real-time reasoning. In: Brown, F.M. (ed.) The Frame Problem in Artificial Intelligence, pp. 2349–357. Morgan Kaufmann, Los Altos (1987)
14. Dreyfus, H.: Why Heideggerian AI Failed and How Fixing It Would Require Making It More Heideggerian. MIT Press, Cambridge, Mass (2008)
15. Fitting, M.: A Kripke/Kleene semantics for logic programs. J. Logic Program. **2**, 295–312 (1985)
16. Fodor, J.: The Modularity of Mind. MIT Press, Mass, Cambridge (1983)
17. Gabbay, D.: Theoretical foundations for non-monotonic reasoning in expert systems. In: Apt, K. (ed.) Logics and Models of Concurrent Systems, pp. 439–459. Springer, Berlin (1984)
18. Gallaire, H., Minker, J. (eds.): Logic and Data Bases. Plenum Press, New York (1978)
19. Gelfond, M., Lifschitz, V.: The stable model semantics for logic programming. In: Proc of the 5th. ICLP, pp. 1070–1080. MIT Press, Cambridge, Mass (1988)
20. Gelfond, M., Lifschitz, V.: Logic programs with classical negation. In: Proc of the 7th. ICLP, pp. 579–597. MIT Press, Cambridge, Mass (1990)
21. Goebel, R., Goodwin, S.: Applying theory formation to the planning problem. In: Brown, F.M. (ed.) The Frame Problem in Artificial Intelligence, pp. 207–232. Morgan Kaufmann, Los Altos (1987)
22. Ginsberg, M., Smith, D.: Reasoning abut action I; A possible worlds approach. In: Brown, F.M. (ed.) The Frame Problem in Artificial Intelligence, pp. 233–258. Morgan Kaufmann, Los Altos (1987)
23. Ginsberg, M., Smith, D.: Reasoning abut action II; The qualification problem. In: Brown, F.M. (ed.) The Frame Problem in Artificial Intelligence, pp. 259–287. Morgan Kaufmann, Los Altos (1987)
24. Gryz, J.: The frame problem in artificial intelligence and philosophy. Filozofia Nauki. **21**, 15–30 (1989)
25. Haas, A.: The case for domain-specific frame axioms. In: Brown, F.M. (ed.) The Frame Problem in Artificial Intelligence. Proc. of the 1987 Workshop, pp. 343–348. Morgan Kaufmann, San Mateo, CA (1987)
26. Hanks, S., McDermott, D.: Nonmonotonic logic and temporal projection. Artif. Intell. **33**, 379–412 (1987)
27. Kowalski, R.: Predicate logic as a programming language. In: Proc. of IFIP'74, pp. 569–574 (1974)
28. Kowalski, R.: Logic for Problem Solving. North-Holland, Amsterdam (1979)
29. Kowalski, R., Sadri, F.: Logic programs with exceptions. New Generation Computing
30. Kowalski, R., Sergot, M.: A logic-based calculus of events. New Gener. Comput. **9**, 387–400 (1991)
31. Kraus, S., Lehmann, D., Magidor, M.: Non-monotonic reasoning, preference models and cumulative reasoning. Artif. Intell. **44**, 167–2070 (1990)
32. Kunen, K.: Negation in logic programming. J. Logic Program. **4**, 289–308 (1987)
33. Lehmann, D., Magidor, M.: What does a conditional knowledge base entail? Artif. Intell. **55**(1992), 1–60 (1992)
34. Lukasiewicz, W.: Non-Monotonic Reasoning: Foundation of Commonsense Reasoning. Ellis Horwood, New York (1992)
35. Makinson, D.: General theory of cumulative inference. In: Proc. of the 2nd International Workshop on Non-Monotonic Reasoning, pp. 1–18. Springer, Berlin (1989)
36. Makinson, D.: General patterns in nonmonotonic reasoning. In: Gabbay, D., Hogger, C., Robinson, J.A. (eds.) Handbook of Logic in Artificial Intelligence and Logic Programming, vol. 3, pp. 25–110. Oxford University Press, Oxford (1994)

37. McCarthy, J.: Circumscription-a form of non-monotonic reasoning. Artif. Intell. **13**, 27–39 (1980)
38. McCarthy, J.: Applications of circumscription to formalizing commonsense reasoning. Artif. Intell. **28**, 89–116 (1984)
39. McCarthy, J., Hayes, P.: Some philosophical problems from the standpoint of Artificial Intelligence. In: Meltzer, B., Michie, D. (eds.) Machine Intelligence, vol. 4, pp. 463–502. Edinburgh University Press, Edinburgh (1969)
40. McDermott, D., Doyle, J.: Non-monotonic logic I. Artif. Intell. **13**, 41–72 (1980)
41. McDermott, D.: Nonmonotonic logic II. J. ACM **29**, 33–57 (982)
42. Minker, J. (ed.): Foundations of Deductive Databases and Logic Programming. Morgan Kaufmann, Los Altos (1988)
43. Minsky, M.: A framework for representing knowledge. In: Haugeland, J. (ed.) Mind-Design, pp. 95–128. MIT Press, Cambridge, Mass (1975)
44. Moore, R.: Possible-world semantics for autoepistemic logic. In: Proc. of AAAI Non-Monotonic Reasoning Workshop, pp. 344–354 (1984)
45. Moore, R.: Semantical considerations on nonmonotonic logic. Artif. Intell. **25**, 75–94 (1984)
46. Rolston, D.: Toward a tense-logic based mitigation of the Frame Problem. In: Brown, F.M. (ed.) The Frame Problem in Artificial Intelligence, pp. 319–341. Morgan Kaufmann, Los Altos (1987)
47. Pednault, E.: ADL: Exploring the middle ground between STRIPS and the situation calculus. In: Proc. of the 1st International Conference on Principles of Knowledge Representation and Reasoning (KR'89), pp. 324–332. Morgan Kaufmann (1989)
48. Reiter, R.: A logic for default reasoning. Artif. Intell. **13**, 81–132 (1080)
49. Reiter, R.: Knowledge in Action: Logical Foundations for Specifying and Implementing Dynamical Systems. MIT Press, Mass, Cambridge (2001)
50. Schubert, L.: Monotonic solution of the frame problem in the situation calculus: An efficient method for worlds with fully specified actions. In: Kyburg, H., Loui, R., Carlson, G. (eds.) Knowledge Representation and Defeasible Reasoning, pp. 23–67. Kluwer, Dortrecht (1990)
51. Schwering, C., Lakemeyer, G.: Projection in the epistemic situation calculus with belief conditionals. In: AAAI'15 Proc. of the 29th AAAI Conference on Artificial Intelligence, pp. 1583–1589 (2015)
52. Schwind, C.: Action theory and the frame problem. In: Brown, F.M. (ed.) The Frame Problem in Artificial Intelligence, pp. 121–134. Morgan Kaufmann, Los Altos (1987)
53. Shanahan, M.: Solving the Frame Problem. MIT Press, Cambridge (1997)
54. Shanahan, M.: The Frame Problem. Stanford University, The Stanford Encyclopedia of Philosophy (2016)
55. Shepherdson, J.: Negation in logic programming. In: Minker, J. (ed.) Foundations of Deductive Databases and Logic Programming, pp. 19–88. Morgan Kaufmann, Los Altos (1988)
56. Shoham, Y.: A semantical approach to nonmonotonic logics. In: Proc. of Logics in Computer Science, pp. 275–279 (1987)
57. van Emden, M., Kowalski, R.: The semantics of predicate logic as a programming language. J. ACM **23**, 733–742 (1976)
58. Veltman, F.: Logics for Conditionals. Ph.D.Thesis. University of Amsterdam (1985)
59. Veltman, F.: Defaults in update semantics. J. Philos. Logic **25**, 221–261 (1996)

Chapter 7
Conclusions

Abstract In Chap. 7, we give our conclusions of the book. First, we summarize the results, Second, we show further technical questions which should be solved in our future work.

7.1 Summary and Achievements

In this book we studied knowledge representation and reasoning based on the granular reasoning in the framework of the *epistemic situation calculus* (ES). It properly unifies situation calculus and rough set theory. In our approach, concepts of partiality and overcompleteness play an important role in both knowledge representation and reasoning.

We presented the deduction system for partial semantics utilizing the rough set theory for the basis of knowledge representation and reasoning, and application based on the epistemic situation calculus ES using granular reasoning.

For incomplete and inconsistent information, we proposed the consequence relations and axiomatized systems as three-valued and four-valued logics. Then, we showed the deduction system with Gentzen Sequent Calculi and Semantic Tableaux Calculi.

We axiomatized three-valued and four-valued logics from the partial semantics point of view and showed the relationship between them using consequence relation. Kleene's strong three-valued logic $\mathbf{K}_3$ and the logic of paradox LP by Priest are described with consequence relation and designated value.

$\mathbf{K}_3$ can cope with incompleteness, and LP inconsistency. Using semantic tableaux, we also showed the two types of three-valued logics Kleene strong three-valued logic and the logic of paradox as well.

Chapter 1 gave the motivations and the organization of the present book. Chapter 2 briefly surveys *rough set theory*. Chapters 3 and 4 respectively presented Gentzen systems and tableaux calculi for many-valued logics in connection with rough set theory.

S. Akama et al., *Epistemic Situation Calculus Based on Granular Computing*,
Intelligent Systems Reference Library 239,
https://doi.org/10.1007/978-3-031-28551-6_7

In Chap. 5, we introduced the basic framework to incorporate zooming reasoning based on the granular reasoning into the epistemic situation calculus ES. We used modal logic with four-valued semantics as the interpretation of zooming reasoning.

We also incorporate zooming reasoning into ES as epistemic actions, zooming in and out, which are interpreted as abstraction and refinement of reasoning process respectively. By the four-valued semantics, the system cannot be corrupt when the inconsistent result is deduced.

In Chap. 6, we surveyed and discussed the Frame Problem. We do not have the answer to the Frame problem but we described another point of view by epistemological Frame Problem which is proposed by Dennett.

7.2 Future Direction

We formulated the mechanism of knowledge update by zooming action and sensing in the epistemic situation calculus ES and also to reveal the update process of a knowledge base about the dynamic change of granularity in the model and valuation of information where the zooming action is performed.

We need to describe what is believed after an agent acquires new information by performing an action or by aware of new information from implicit information. Schwering and Lakemeyer [4] studied conditional belief for the preferential belief structure.

Beliefs about different contingencies are expressed through if-then statements so-called *conditional beliefs*. The preferential belief structure is initially determined using conditional statements.

For an intelligent agent–a robot, for example–it is important to behave reasonably in such dynamic and uncertain conditions. The key problem that arises when reasoning about actions and beliefs is the belief projection problem: Schwering et al. investigate what the agent believes after a sequence of actions brings about physical or epistemic change.

Physical and epistemic changes are caused by actions. Physical change means that an action may directly affect and change the value of predicates. Epistemic change, on the other hand, means that an action may convey new information that leads to a reassessment of how plausible the agent considers its different beliefs to be, namely this information is taken into account by belief revision.

Murai et al. [1] studied the incomplete and inconsistent feature of conditional logic and their measure-based extensions based on granular reasoning in the framework of conditional models. Then paracomplete and paraconsistent aspects of conditionals are examined in the framework.

The conditional logic with measure-based extensions is suitable to capture the incomplete and inconsistent aspects of initial and updated knowledge state using conditional belief. We need the semantic model for the conditional belief with extended conditional logic and will be left to our future works.

Proof methods for bilattice logics need a work. Possible methods include natural deduction, sequent calculi and tableaux calculi, which contribute to decision logic. Initial works for this direction were recently done by Nakayama et al. [2, 3].

Furthermore, in another direction, we will study intention in the framework of the epistemic situation calculus. In the situation calculus, the intention is not treated clearly, but the meta predicate G of projection of the situation calculus can be interpreted as an action for the intention. Therefore, we will treat the intention and interpret it as an epistemic action and another modal operator.

As discussed in the previous chapter, the epistemological Frame Problem seems to be interesting for the formalization of common-sense reasoning in AI. Because our epistemic situation calculus addresses the roughness of our knowledge, it may be able to shed new light on the epistemological Frame Problem in the context our cognitive activity.

References

1. Murai, T., Kudo, Y., Akama, S.: Paraconsistency, Chellas's conditional logics, and association rules. In: Abe, J.M. (ed.) Towards Paraconsistent Engineering, pp. 179–196. Springer, Heidelberg (2016)
2. Nakayama, Y., Akama, S., Murai, T.: Billatice logic for rough sets. JACII **24**, 774–784 (2020)
3. Nakayama, Y., Akama, S., Murai, T.: Many-valued tableau calculi for decision logic based on approximation regions in VPRS. J. Reasoning-based Intell. Syst. **13**, 235–242 (2021)
4. Schwering, C., Lakemeyer, G.: Projection in the epistemic situation calculus with belief conditionals. In: AAAI'15 Proc. of the 29th AAAI Conference on Artificial Intelligence, pp. 1583–1589 (2015)
5. Shanahan, M.: The Frame Problem. Stanford University, The Stanford Encyclopedia of Philosophy (2016)

Index

S. Akama et al., *Epistemic Situation Calculus Based on Granular Computing*, Intelligent Systems Reference Library 239,
https://doi.org/10.1007/978-3-031-28551-6

Printed in the United States
by Baker & Taylor Publisher Services